ADVANCED GENETIC ANALYSIS

Finding Meaning in a Genome

To my Father, in memoriam, and to Jim and Virginia Stowers whose "Hope for Life" inspired the completion of this book. RSH

To Dr. Michael Abruzzo – for introducing me to the science of Genetics properly; and to Dr. Stephen D. Phinney – for believing before anyone else did, and for demonstrating what it's like to think outside the box. MYW

And for the monotremes everywhere . . .

Advanced Genetic Analysis

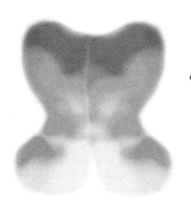

Finding Meaning in a Genome

R. Scott Hawley
Stowers Institute for Medical Research

and

Michelle Y. Walker
Galileo Laboratories Inc.

Blackwell
Publishing

© 2003 by Blackwell Science Ltd

BLACKWELL PUBLISHING
350 Main Street, Malden, MA 02148-5020, USA
9600 Garsington Road, Oxford OX4 2DQ, UK
550 Swanston Street, Carlton, Victoria 3053, Australia

First published 2003 by Blackwell Science Ltd, a Blackwell Publishing company

4   2007

Library of Congress Cataloging-in-Publication Data

Hawley, R. Scott.
  Advanced genetic analysis : finding meaning in a genome / R. Scott Hawley and Michelle Y. Walker.
    p. cm.
  Includes bibliographical references and index.
  ISBN 1-4051-0336-1 (pbk. : alk. paper)
  1. Genetics—Research—Methodology. I. Walker, Michelle Y. II. Title.

QH440 .H39 2002
576.5'072—dc21

                                                         2002071233

ISBN-13: 978-1-4051-0336-7 (pbk. : alk. paper)

A catalogue record for this title is available from the British Library.

Set in 9.5/12pt Minion
by Graphicraft Ltd, Hong Kong
Printed and bound in the United Kingdom
by TJ International Ltd, Padstow, Cornwall

The publisher's policy is to use permanent paper from mills that operate a sustainable forestry policy, and which has been manufactured from pulp processed using acid-free and elementary chlorine-free practices. Furthermore, the publisher ensures that the text paper and cover board used have met acceptable environmental accreditation standards.

For further information on
Blackwell Publishing, visit our website:
www.blackwellpublishing.com

Contents

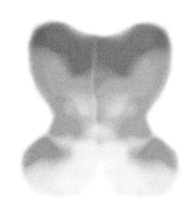

Contents

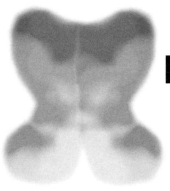

Preface

"For the geneticist there are accordingly three ways of examining anything. Through characters, he can examine function; through their changes, he can examine mutation; through their reassortment, he can examine recombination."

Francois Jacob in *The Logic of Life* (p. 224)

This book is intended for an advanced course in genetic analysis. The focus is on the basic principles that underlie genetic analysis: mutation, complementation (and its bridesmaids, suppression and enhancement), recombination, segregation, and regulation. Our goal is to provide insights into the biological and analytical processes that comprise each of these tools, and to explain their use. Our basic objective is to explain to you just what each of these tools, or operations, does, and how that operation or test can lie to you. Perhaps most importantly, the book is designed to teach you just how much you can learn when nature misunderstands your question. In other words, this is a book about genetic theory.

Although a discussion of genetic analysis inevitably requires the presentation of multiple examples, this is not to be considered as a textbook of genetic facts. Facts can sometimes change in the blink of an eye; but the basic analytical tools change rather more slowly. We have tried to be as comprehensive and catholic as possible in the choice of examples and organisms, drawing on examples from as many genetically tractable model organisms as possible, with even the occasional reference to humans. The book assumes the reader has a basic familiarity with the genetics of eukaryotes, as well as with the basic biology of prokaryotes and their viruses. We also take for granted a working knowledge of the three sirens of molecular biology: transcription, translation, and replication. In cases where specialized techniques are used or concepts required, we have endeavored to provide the essential background material.

We are obviously not the first authors to attempt to set forth the principles of genetic analysis. Indeed, this book follows in the footsteps of Pontecorvo's *Trends in Genetic Analysis* (Pontecorvo 1958) and Stahl's *The Mechanics of Inheritance*, 2nd edn. (Stahl 1969). We have also drawn heavily from courses taught by Professor Dean Parker, Professor Kenneth W. Cooper, and the late Professor Crellin Pauling at UC Riverside, and from courses taught by Drs. Jonathan Gallant, Leland Hartwell, and the late Drs. Larry Sandler and Herschel Roman at the University of Washington. RSH also acknowledges gratefully a

20-year on-going and intensive genetics tutorial with Professor Adelaide Carpenter at Cambridge.

This book had its beginnings while the authors were members of the Section of Molecular and Cellular Biology at the University of California at Davis. The support of the faculty and staff of that Section is gratefully recognized. The book was completed after RSH moved to the Stowers Institute for Medical Research in Kansas City. RSH gratefully acknowledges the support of many members of the Institute, and especially Dr. Bill Neaves, was critical to the completion of this project. His gratitude extends to Jim and Virginia Stowers, whose vision of "Hope for Life" extends far beyond the Institute they founded.

We are especially indebted to our editors at Blackwell for help during all phases of this project, but most especially for knowing when to prod and when to let us be – to Nancy Whilton and Nathan Brown, we offer a most sincere "thank you." Nancy showed a calmness that allowed us to respond to criticism in a constructive rather than a reactive fashion. RSH will long view the lessons learned from her to be the major reward of writing this book. Nathan was patient to the point of stoic and dedicated to the point of heroic.

Many people read and extensively commented on various drafts; we especially thank Professors Frank Stahl, Jim Haber, and Julia Richards. These three colleagues went far beyond anything we might have expected to help make this a "useful book." The debt to Julia Richards for thoroughly revising the section on LOD score analysis is enormous and the gratitude equally large. Our colleagues Mike Zwick, Kent Golic, Ken Burtis, Keith Maggert, Pam Geyer, Ting Wu, Mia Champion, and Kim McKim also read chapters and made important contributions. We also thank Dan Kiehart, Tim Stearns, Adelaide Carpenter, Tom Cline, Seymour Benzer, Carl Marrs, Michael Dahmus, and many others who took the time to answer questions or explain techniques. We thank David Harris, Charisse Orme, and Heather Peters for insightful comments and for proofreading the entire text. We owe a real debt to Ed van Veen for teaching us the math we should have known and for editing, re-editing, and sometimes rewriting much of chapter 7. A number of "outside reviewers" contributed significantly to the clarity and accuracy of the text: Celeste Berg, University of Washington; YiPing Chen, Tulane University; Todd Disotell, NYU; Richard Gaber, Northwestern University; Kim McKim, Rutgers University; John Osterman, University of Nebraska at Lincoln; Paul Ramsey, Louisiana Tech.

RSH also acknowledges a debt to Diana Hiebert and Nancy Lane for secretarial help, and for preserving his sanity. In that vein, he also, once again, acknowledges the muses Jewel, Harriet Schock, Meat Loaf, John Stewart, Trisha Yearwood, and Pam Tillis for keeping him company during much of the writing. RSH thanks his teenage daughter, Tara Hawley, for her assistance in creating the index.

The inception for this book was a course in Advanced Genetic Analysis taught by RSH at UC Davis. A large debt is owed to the students, and most especially the Teaching Assistants, of that course. We especially thank Shelley Force, Chris Boulton, Charisse Orme, Heather Peters, David Harris, Kara Koehler, Mike Zwick, Rachael French, Soni Lacefield, and Bill Gilliland for raising questions that were the impetus for much of the book. Every point of clarity in this book came from wrestling with a question they asked. More than likely, it was answered badly (or gruffly) at the time. But it wasn't forgotten. The need to provide cogent answers to those questions rumbled about in our heads for years. Our effort to answer them properly is the book that is now in your hands.

Finally, the last section of this tome is a list of some 430 references. Despite every attempt to be complete, we know that we have missed citing many extremely well-done and important examples of genetic analysis. In our defense, we can say only that there came a time to stop reading and to actually write this book. Our apologies to those authors whose work was in the huge stack of "you know we really ought to discuss this" papers on the side of our desk.

"A genetic unit should be defined by a genetic experiment. The absurdity of doing otherwise can be seen by imagining a biochemist describing an enzyme as that which is made by a gene."

Seymour Benzer, *Scientific American*

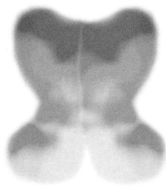

Introduction

"All in due time, my pretties, all in due time."
The Wicked Witch of the West

One either begins a genetics textbook by talking about Gregor Mendel or about James Watson and Francis Crick. That is simply the way things are done. The choice an author (or authors in this case) makes reflects their basic scientific predilections. Classical geneticists start with Mendel, and thus so will we. The work of Drs. Watson and Crick will be considered in due time.

To Gregor Mendel, a gene was little more than a statistical entity, defined by effects on phenotypic variation and by segregation. Each trait was determined by two copies of a given gene. Differences in the trait (phenotype) were due to differences in the "form" of a given gene. These different forms of a given gene were called alleles. In Mendel's construction, the phenotype was a direct consequence of the alleles present in a given individual. To explain the cases where the effect of one allele predominated over the other, Mendel created the concepts of dominance and recessivity. Thus, to Mendel, the primary definition of a gene was that it was a unit of hereditary information. Genes were not the structures or tissues themselves, rather they provided information required to create those structures.

Mendel's concept of the gene was also firmly embedded in the idea of a gene as the unit of segregation. Each individual possesses two copies of each gene, one copy inherited from the mother and one copy inherited from the father. Moreover, each individual passes on only one of those two genes to their own offspring, and they do so at random. The gene pair thus becomes the unit of segregation and that process of segregation ultimately leads to gametes bearing a single hereditary particle (gene) for each trait. Mendel's concept of independent assortment can be thought of as a rather simple extension of this idea, i.e. because the individual gene pair is the unit of segregation, the assortment of two gene pairs will occur at random.

So Mendel's concept of the gene, as we now understand it, was all but Newtonian. Mendel saw genes as small immutable particles whose movement was controlled by natural law in such a way that it could be modeled statistically. We can describe Mendel's concept of the gene in three laws. The first law, which we will call the "Purity and Constancy of the Gene," states that genes themselves are immutable. Although genes produce the phenotype, they are not themselves the phenotype. Nor are they affected by the phenotype. The second law, the "Law of the Gene," states that each individual carries two, and only two, copies of a gene for a given trait. This law also states that each time a gamete, a sperm or an egg, is made, one of those two copies is chosen at random to be included in that gamete. The third law, the "Law of Independent Assortment," mandates that for two pairs of genes, the choice of which copy of gene A is included in the gamete does not affect the choice of which copy of gene B is included in the same gamete.

The only real omission in Mendel's analysis resulted from his lack of understanding of the gene itself. Indeed, the first major advance in genetical thinking would focus on correcting Mendel's view of the gene as an immutable object; his idea of the purity and constancy of a gene. Genes might be very constant, and indeed they are, but they would not turn out to be immutable. Mutations in many, but perhaps not all, genes can produce some phenotypic effects. Moreover, the modern geneticist is savvy enough to know that most biological processes involve the products of multiple genes. Thus, for at least many traits, there may be multiple genes that can mutate to a given phenotype. We are also aware that for some fraction of genes the connection between genotype and phenotype may be influenced both by other genes and by the environment.

We have come a long way since Gregor Mendel. We have a much clearer view of the gene. Along the way we have developed some very impressive tools to study gene function. These are the focus of this book. But before we embark on our discussion, perhaps we should ask ourselves "why should we bother with doing 'genetics' at all?" What can we obtain from the isolation and analysis of mutants (for that is what genetics is) that we cannot learn by one of the "-omics" of modern biology?

We will define genetics simply. *Genetics is both the use of mutations and mutational analysis to study a given biological process and the study of the hereditary process itself*. When done right, the two halves of that description are inextricable. Thus if someone is isolating mutants and characterizing those mutants to study flight, they are doing genetics. If they are simply isolating genes expressed in bird muscle, they may be doing biochemistry, but they are not doing genetics. *The very core of genetics is mutation*. However, the actual doing of genetics requires more than isolating mutations. Doing genetics well also requires that investigators isolate and characterize those mutants in a fashion that: (a) maximizes their chance of answering their initial questions; (b) provides them with as many novel biological insights as possible; and (c) facilitates a greater understanding of the structure and function of the genome they are studying. In other words, "doing genetics" well means understanding what types of mutations one can get, how to get them, and how to analyze them. For example, the analysis of suppressor mutants can be a powerful tool indeed when done correctly.

The proper *doing of genetics* requires that one understands one's tools. The basic intellectual tools of genetics are: mutation, complementation, suppression, recombination, and regulation (epistasis). This book is about the proper use of those intellectual tools. Our goal is to give you ideas about what works, and cautions about what doesn't. We will discuss the biology and biochemistry of many processes, but only when we need to do so to describe the mutants. The very essence of our story is the mutants. So that is where we begin. . . .

Mutation

A mutation is a stable and heritable change in a DNA sequence. Mutations that occur within or near a gene often create a phenotype different from that normally expressed by the wildtype allele of that gene. A number of different types of mutations have been found to cause changes in phenotype. These include base substitutions (e.g. C → T), duplications, or deletion of base pairs, and various chromosomal aberrations such as inversions and translocations. Whether it changes a single base pair or rearranges entire chromosomes, a heritable change in the base pair sequence of a cell's DNA is considered a mutation. Because this book is fundamentally about mutational analysis, we need to spend some time considering the types of mutations that can occur, both at the molecular level and in terms of the effects they have on phenotype. We also need to review the various systems that exist to classify mutations. Such a review is all the more critical because the nomenclature systems that geneticists have developed are keyed to the structure and effects of the mutants they name. Thus, the things themselves, and our names for them, are inextricably intertwined.

1.1 Types of mutations

Most introductory genetics texts classify mutations simply as recessive or dominant. A mutation (*m1*) is said to be **recessive** if *m1/m1* organisms display a mutant phenotype, but +/*m1* organisms are wildtype. (*Note on symbolism*: the symbol "+" denotes the wildtype, or normal, allele of a given gene.) Conversely, a mutation (*M2*) is said to be **dominant** if *M2*/+ organisms display a mutant phenotype while +/+ organisms are normal. Some texts use the term **semidominant** to describe cases where a dominant mutation, *M3*, displays a more severe (or extreme) phenotype as a homozygote (*M3/M3*) than it does as a heterozygote (*M3*/+), such that the order of phenotypic severity is *M3/M3* > *M3*/+ > +/+. While such a classification is sufficient for some purposes, it is inadequate to actually describe the range of mutant types that can be observed. Accordingly, at least three more detailed classification systems have been developed, and these are discussed below.[1]

recessive   m1/+ : w.T
dominant   m2/+ : Δ
semidominant m3/m3: more
                        severe

---

[1] Intertwined with the issue of mutant classification is the problem of genetic nomenclature. Unfortunately, each organism uses a different system to symbolize gene names. The most thorough comparative summary of the existing nomenclature systems for various model organisms is published in a special issue of *Trends in Genetics* (March, 1995). Also included in this chapter are boxes describing each of the four model organisms that will be discussed often in the text. Brief summaries of their nomenclature systems, and references to more detailed nomenclature guides, are provided in these boxes.

1.1.1   Muller's classification of mutants

The first detailed mutant classification scheme was proposed in 1932 by H. J. Muller. Muller (1932) classified mutations into five basic groups: nullomorphs, hypomorphs, hypermorphs, antimorphs, and neomorphs. The assignment of a mutant to one of these classes was largely based on Muller's view that mutations can, and should, be described in terms of their effect on activity. A mutation can be assigned to one of these five groups by comparing the phenotypic effects of that mutation in homo-, hetero-, and hemizygotes. Understanding these classifications, and being able to use them, is a critical component of genetic analysis; therefore we will consider each of these types of mutations in some detail. We begin by considering the two classes of loss-of-function mutations: nullomorphs and hypomorphs.

Nullomorphs

Nullomorphs (or amorphs) are mutants with no remaining gene function, i.e. they produce no functional gene product. They are often, and far more precisely, called null alleles, and are the basic mainstay of genetic analysis. Nullomorphic mutations might correspond to internal deletions, frameshift mutations that occur early in the coding sequences, or missense mutations that alter a critical site in the protein in such a way as to fully ablate its activity. The most characteristic feature of a nullomorph is that it is the equivalent of a full deletion of the gene in terms of its influence on the final phenotype.

Null alleles lead to the complete absence of a functional protein product via a variety of defects in gene expression. Once the relevant molecular tools are available, one can, and usually does, discriminate between transcriptional nulls, protein (or translational) nulls, and mutations that produce completely inactive proteins. In the case of a transcriptional null, no full-length transcript is produced. Such mutations might reflect, for example, the deletion of crucial elements in the promoter. "Protein nulls" are defined simply as mutations that fail to produce a protein product, as assayed by an antibody specific for that protein. (This classification may include transcriptional nulls in cases where the mutant's effect on transcription has not been assessed.) Inactivating nulls produce a protein product, but that product exerts no obvious activity. However, the most obvious type of nullomorphic mutation is the deficiency (or deletion).

According to Muller, most mutations "involve more or less the inactivation of the processes governed by the normal gene, and . . . these less active genes should more often act as recessives." In the case of genes whose products are enzymes, one can easily understand why most loss-of-function mutations, even nullomorphs, are recessive. For most enzymes the reaction rate *in vivo* is limited by substrate concentration, so that reducing the concentration of the enzyme by a factor of two (as one presumably does in m/+ heterozygotes) may be expected to have little effect on the rate of product formation.

However, not all nullomorphic mutations will be recessive. Nullomorphic mutations in genes that code for structural proteins, or for enzymes whose function is highly concentration-dependent, may well be dominant. Some of the *Minute* mutants of Drosophila are excellent examples of nullomorphic mutants that can exert dominant effects. *Minute*/+ flies show delayed development and a variety of morphological anomalies, including short, thin bristles. *Minute*/+ females show reduced fertility or, in extreme cases, are sterile. *Minute*/*Minute* progeny are inviable. Most strong *Minute* mutations are due to loss-of-function mutations in genes that encode ribosomal proteins. The reduction of the quantity of even a single ribosomal protein by a factor of two apparently reduces twofold the final number of ribosomes in the cell as well. The consequence of a twofold reduction in the number of ribosomes is a corresponding decrease in protein synthesis capacity, which has dramatic consequences for the phenotype of the organism.

Hypomorphs

Hypomorphs are mutants that produce some degree of residual activity, but not enough to provide wildtype activity in *m/m* homozygotes. Indeed, the term hypomorph is often synonymous with "weak allele." Hypomorphic mutations reduce the amount or level of activity of the protein product. One can imagine a host of genetic lesions that might produce hypomorphic mutants, ranging from mutations that decrease the level of transcription to mutations that alter the messenger stability, or the activity or amount of protein product. None the less, the defining characteristic of a hypomorphic mutant is that some discernable level of active product is being produced. Most hypomorphic mutants are recessive, but the same caveat about loss-of-function mutations in genes encoding proteins whose dosage is critical applies to partial loss-of-function mutants as well.

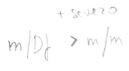

One can easily distinguish between hypomorphic and nullomorphic mutants if a deficiency (*Df*) for the gene in question is available. A deficiency is a deletion of DNA from the chromosome that encompasses the gene in question (figure 1.1). For a hypomorph, *m/Df* is expected to be more severe in phenotype than an *m/m* homozygote. (Taking into consideration that the *m/m* homozygote is expected to produce twice as much active product as the *m/Df* individual.) A true nullomorph is the genetic equivalent of a deficiency; thus in terms of phenotypic severity, *m/Df* and *m/m* are expected to be equivalent. By a similar calculation, an individual with three doses of a hypomorphic mutation (*m/m/m*) is expected to be less severe than one with two doses (*m/m*). Adding more doses of a true nullomorphic allele should have no effect.

We have heard many geneticists refer to hypomorphs as the bane of their existence. True enough, the residual level of activity created by such mutations often

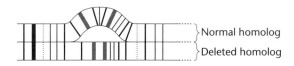

> Normal homolog
> Deleted homolog

Figure 1.1 A deficiency heterozygote as revealed by polytene chromosome analysis

frustrates the phenotypic or functional analysis of these genes. However, hypo-morphic mutants are often the first or only mutants to define important genes. It turns out, as we shall discuss in the next chapter, that many mutant hunts or screens require homozygotes for the newly induced mutations to be viable. Given this restriction, a null allele of a gene required for life would not be recovered in such a screen, even if the protein product of that gene played a critical role in the process under study. In contrast, a hypomorphic mutant can, and often does, give enough product to allow survival, but not enough to pro-duce a normal phenotype. In such a case, the finding of a hypomorph alerts the investigator to the existence of this gene, and heralds its role in the process under study.

Hypermorphs

As the name implies, hypermorphs produce either a hyperactive protein product or a harmful excess of the normal protein product. The defining characteristic of this mutant class is that m/Df should be less severe in phenotype than $m/+$. Indeed, in terms of decreasing phenotypic severity, the dosage series should be $m/m > m/+ > m/Df$. Verified examples of hypermorphic mutations are few and far between. The best example of a hypermorphic mutation in Drosophila is a mutant called *Confluens* that affects wing vein morphology. *Confluens* is an allele of the *Notch* (N) gene. The phenotype of *Confluens* can be mimicked by three doses of $N+$, and *Confluens* over a nullomorphic allele of *Notch* is wildtype. All of this makes perfect sense when one realizes that the *Confluens* mutation is a tan-dem duplication of the $N+$ gene. Clearly, increases in the dose of the $N+$ gene have phenotypic consequences, and however one gets to three doses, one creates a phenotype.

One can imagine other types of mutations that might up-regulate the tran-scription or translation of a given gene or its mRNA product, and thus produce observable phenotypes. It is perhaps surprising that so few bona fide hyper-morphs have been described. However, a new technique for creating hyper-morphs in Drosophila has been developed by Pernille Rorth (Rorth et al. 1998). Rorth began by creating a transposon, or mobile element, that was capable of inducing genes neighboring that transposon to be highly expressed in a fashion that was both tissue-specific and inducible. By mobilizing that transposon within the Drosophila genome, Rorth and her collaborators created a collection of 2,300 lines of flies, each of which carried an independent transposon insertion. This collection of insertions was screened for the ability to suppress a hypomor-phic mutation in the *slow border cells* (*slbo*) gene that confers a cell migration defect. This defect is observed in the Drosophila ovary and results in sterility. Because the *slbo* gene encodes a C/EBP transcription factor, Rorth and her col-laborators reasoned that the high-expression "suppressors could be genes nor-mally activated by C/EBP in border cells, or genes which (in this situation) are rate-limiting for cell migration". They obtained both.

Of 2,082 insertion lines tested, 60 showed clear suppression of the mutant phenotype created by homozygosity for *slbo*. The suppressing insertions resulted in the over-expression of genes that encoded known players in actin cytoskeletal remodeling, a critical process in cell migration. They also recovered insertions in a receptor tyrosine kinase gene (*abl*) that appears to be involved in the control of actin polymerization. The success of the Rorth suppression screen may have been largely due to the choice of a hypomorphic *slbo* allele that retains some degree of C/EBP activity. Thus, it was only necessary to provide a small increase in border cell migration to suppress the sterility caused by the *slbo* mutation.

If they can be obtained, hypermorphs can be valuable tools for dissecting a genetic process. This may be especially true in the case where one is dealing with a group of functionally redundant genes. In such a case, a simple loss-of-function mutation in one of these genes may not produce a discernable phenotype, but over-expressing one of the genes may create an observable defect. Indeed, we will consider the use of high-copy suppression libraries in yeast to mimic the creation of hypomorphic mutations, by creating colonies each of which possess a high-copy number of a plasmid carrying a given gene. Phenotypes created by such methods can also serve as the substrate for "enhancer" and "suppressor" screens aimed at identifying other genes in this process.

Antimorphs

An antimorphic mutation results in a protein product that antagonizes, or poisons, the wildtype protein. Thus the phenotype of a true antimorph is expected to mimic the phenotype presented by a strong hypo- or nullomorph. Antimorphic mutations are dominant, by definition. However, increasing the dose of the wildtype allele can often ameliorate the phenotype of an antimorphic mutant. For example, imagine an enzyme comprised of four identical subunits of protein A that is required to synthesize the chemical "muctin." Nullomorphic mutations in gene A should produce a muctin-minus phenotype when homozygous. Assuming that half the level of the enzyme is enough, the +/*a* heterozygotes should be normal. Loss-of-function alleles of gene A will be recessive. But now imagine a mutant allele of gene A that produces an unusual structural variant of protein A. This variant protein A is incorporated into the polyprotein complex in such a way as to render the entire complex inactive. Assuming that the wildtype and mutant protein A molecules are produced with equal abundance, this variant of protein A will inactivate virtually all (15/16) of the polyprotein complexes, and the remaining activity may simply not be enough. However, by increasing the dose of the wildtype allele to two, approximately 20% of the polyprotein complexes will be comprised of normal subunits even in the presence of one copy of the antimorphic allele.

A more vivid (and real) example is presented in figure 1.2. The microtubules that make possible many processes of cellular movement are comprised of long arrays of tubulin monomers. Each of these monomers is composed of an alpha-tubulin and a beta-tubulin subunit. Imagine a mutant in the beta-tubulin gene that produces a variant subunit that can incorporate into a growing chain, but cannot support further growth. Once a mutant subunit is incorporated, chain growth freezes. However, by increasing the dosage of the normal allele, one

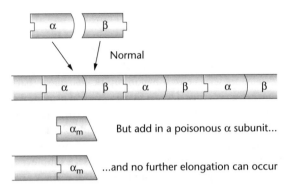

Figure 1.2 Incorporation of the product of an antimorphic allele of tubulin impedes further growth on the microtubule fiber

decreases the probability of incorporating a mutant subunit from one-half to one-third. One *might* see some phenotypic amelioration in such a case. Indeed, a dominant mutant ($\beta 2t^D$) allele of a testis-specific beta-tubulin gene has exactly this type of effect on microtubule assembly in the male germline of Drosophila. Both heterozygotes and homozygotes show dramatic defects in the formation of large microtubule assemblies in the testis, and are sterile as males. In contrast, heterozygotes carrying an extra dose of the wildtype allele ($\beta 2t^D/+/+$) are weakly fertile (Kemphues et al. 1980, 1983).

As noted above, the vast majority of antimorphs are dominant. If we use the symbol A to denote the mutant, the relative phenotypic severity observed in different genotypes can be described as follows:

$A/A \geq A/Df \geq A/+ >>> +/Df \geq +/+$

The defining characteristic of an antimorph is that one should be able to revert (or more precisely "**pseudo-revert**") an antimorphic mutant to a nullomorphic mutation of the gene. In other words, the easiest way to stop this allele from producing a poisonous product is to simply inactivate the gene by a second intragenic mutation (e.g. any mutation that blocks production of the poisonous protein). The result should be a new loss-of-function, preferably nullomorphic, allele, denoted r^A. Most critically, the following should be true:

$+/r^A$ individuals should be phenotypically similar to $+/Df$ or $+/+$ individuals (i.e. the original antimorph should be reverted)

and

r^A/r^A individuals should be phenotypically similar to A/A, A/Df, or $A/+$ individuals

Indeed, the ($\beta 2t^D$) allele generated by Kemphues can be reverted to create recessive loss of function alleles of the ($\beta 2t^D$) gene that follow the rules just set forth (Kemphues et al. 1983).

As a second example of the processes of mutating an antimorph into a nullomorph, consider the Drosophila gene *nod*. The *nod* gene is required for female meiosis in Drosophila, but it is also expressed in virtually all mitotically dividing cells (Zhang et al. 1990). There are many recessive loss-of-function *nod* alleles that disrupt meiotic chromosome segregation when homozygous. Curiously, none of these mutations has any demonstrable effect on mitotic cell division or mitotic cells. There is, however, one dominant allele called *nod^{DTW}*. Heterozygous *nod^{DTW}*/+ females show the same defect in chromosome segregation as do females homozygous for complete loss-of-function *nod* mutations (Rasooly et al. 1991).

The normal function of the Nod protein is to stabilize chromosomes along microtubule tracks. The *nod^{DTW}* mutation poisons that process, and appears to lock the chromosomes in place. The *nod^{DTW}* mutation alters only a single amino acid in a critical region of the protein. Rasooly et al. (1991) mutagenized males carrying the *nod^{DTW}* mutation on their X chromosomes and screened for "pseudo-revertants" (i.e. mutated X chromosomes that no longer exhibited the dominant meiotic effect). They recovered four such mutants, all of which turned out to be new nullomorphic and fully recessive alleles of *nod*. When sequenced, each of these new mutants carried the original *nod^{DTW}* mutation as well as a second mutation that inactivated the *nod* gene.

Neomorphs

It is not clear exactly what Muller intended by this term. But over the years, neomorph has come to mean a mutation that causes a gene to be active in an abnormal time or place. One of the best examples is a somatically arising translocation event that occurs in human beings and results in a blood cancer known as Burkitt's lymphoma. As a result of a translocation between chromosomes 8 and 14, the coding sequence of the *myc* gene on chromosome 8, which acts to *promote* cell division, now lies downstream from a very powerful set of lymphocyte-specific promoter elements derived from a gene on chromosome 14. After translocation, these promoter elements inappropriately turn on the *abl* gene in white blood cells, resulting in uncontrolled cellular proliferation.

Alternatively, one can consider a dominant mutation (*Antp^{73b}*) in Drosophila that causes the antennae to be replaced by legs. (That is, there are two extra legs sticking out of the head just above the eyes.) As diagrammed in figure 1.3, this mutation results from an inversion that places some of the coding sequences of the normal *Antennapedia (Antp)* gene next to the 5′ coding region of a gene normally expressed in the head. Note that the *Antennapedia* gene is normally not expressed in the head. Rather, Antennapedia is expressed in the thorax of the developing embryo, and plays a critical role in specifying the development of thoracic structures such as the leg. But as a result of this inversion, the 3′ region of the *Antennapedia* gene has been fused to the 5′ end of a head-specific gene. As a result of this fusion, a significant portion of the Antennapedia protein, enough to specify leg development, is expressed in the antennae primordia of the developing head. The result is legs where there should be antennae.

Finally, a set of recent studies by Ganetzky and colleagues has served to elucidate the identity and function of one of the most fascinating of all neomorphic

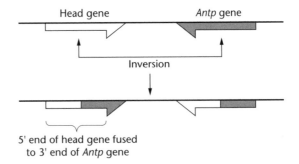

Figure 1.3   *Antp^{73b}* results from an inversion that fuses the 3′ coding sequences of the *Antp* gene with the 5′ coding sequences and regulatory sequences of a gene normally expressed in the head. (From Frischer et al. 1986)

mutants, the *Segregation Distorter* (*Sd*) chromosome in Drosophila (Merrill et al. 1999; Kusano et al. 2001a,b). In the male germline of Drosophila, the *SD* chromosome exhibits a process referred to as "meiotic drive." When heterozygous with a normal chromosome, *SD* chromosomes have the endearing habit of destroying their wildtype homologs (denoted *SD^+*). 99% of the sperm produced by *SD/SD^+* heterozygote will carry the *SD* chromosome and less than 1% will carry the wildtype homolog. The *SD* chromosome does this by causing its *SD^+* brethren sperm to destroy themselves post-meiotically during the process of spermatid development.

An *SD* chromosome is composed of several genetic units that contribute to its function. The first of these is the *Sd* mutant itself, which can act on a separate second chromosomal target element called *Rsp* to cause spermatid dysfunction (Ganetzky 1977). *Rsp* itself is comprised of a repetitive element located in the centric heterochromatin of the second chromosome (Wu et al. 1988). The sensitivity of a given chromosome to destruction by an active *Sd* element increases with the number of copies of the *Rsp* repeat. The *Sd* mutant is the result of a small tandem duplication event involving the RanGAP gene (Powers & Ganetzky 1991) that results in a mutant RanGAP protein, truncated by 234 amino acids at the C-terminus (Merrill et al. 1999). This truncation creates a novel protein whose expression causes distortion – it is the presence of this novel form of the protein gives rise to meiotic drive. Loss-of-function mutations in the wildtype (*SD^+*) RanGAP gene are not expected to create an *SD* phenotype.

There is an important lesson here. Early papers on *SD* often focused their efforts on the idea that understanding the mechanism by which *SD* induced distortion might provide critical insights into the meiotic mechanism itself. The reality is that the actual function of the RanGAP gene is unrelated to the mechanism of maintaining Mendelian fairness, rather it is a function required to mediate nuclear transport in many if not most cell types. However, it was a mechanism that lent itself to exploitation in a wonderfully devious way. H. J. Muller was clearly prescient in putting forward the idea of a *neomorph*, a mutation that creates a novel phenotype function unrelated to the usual function of the gene.

One major distinction between antimorphs and neomorphs is that the neomorph is **not** poisoning the normal function of the unmutated gene, it is rather performing the correct function at the wrong time or place. Operationally, this distinction can be made by attempting to pseudo-revert the dominant mutation. Reversion of the Ant^{73b} mutation often results in a complete null mutation in the *Antennapedia* gene. The phenotype of that mutation is early embryonic lethality due to malformations of the thorax. Compare this to the *nod* antimorph described earlier, where the original antimorph and its loss-of-function revertants had the same phenotype. Neomorphs also differ from antimorphs in that the effects of neomorphic alleles are independent of the presence or dosage of the wildtype allele. Thus, by definition, neomorphic alleles are dominant.

In recent years the term neomorph has gradually been supplanted by the term "gain-of-function mutation" (see below). Actually, both can be somewhat misleading terms, because in most cases the mutant does not confer a truly "new" or "gained" function on the gene; rather it simply causes the gene to be expressed at the wrong time or in the wrong place. None the less, it seems odd to us that there is not even a single example that fits the definition of a neomorph, that is to say a mutation that creates a protein with a "novel" function.

1.1.2 *Modern mutant terminology*

Following the elucidation of the central dogma of molecular biology (DNA → RNA → protein), some geneticists felt that Muller's classically based system was too awkward to describe the alterations in gene function created by mutants at the molecular level. In modern articles, mutants that reduce the level of gene product are often classified only as **loss-of-function** mutants. This term often lumps together hypomorphic and nullomorphic mutants. More precisely, loss-of-function mutants can be characterized further, as follows.

- *Null mutants*: complete loss-of-function mutants. Indeed, in modern parlance the term "null mutant" is reserved for those cases where the molecular biology of the mutant gene is well enough understood for us to be confident that there is **no functional** gene product produced. Often the definition of this term is narrowed further to imply that no product is produced at all.

- *Partial loss-of-function mutants*: mutants where some degree of product activity, or product itself, is produced by the mutant protein. We should realize that partial loss-of-function mutants include both those mutants that impair the level of product formation and those that create a partially functional product.

- *Temperature-sensitive (or other conditional) loss-of-function mutants*: the term refers to cases where the loss of activity of the gene protein is observed under one set of conditions (i.e. at higher temperature) and not under the other. The canonical case of a temperature-sensitive mutant involves the denaturation of the mutant protein at higher temperatures. This is not, however, always the case. In some circumstances even null mutants can be sensitive to environmental cues, and produce a phenotype only in the presence of an environmental or genetic stress.

Dominant mutants are referred to by several different names, some of which seem more or less useful than their Mullerian counterparts. For example, mutants that produce poisonous products are referred to as ***dominant negative*** mutants. This is a well-used and often accurate term, but we fail to see its advantage over "antimorph." Mutants that cause a gene to be inappropriately expressed during development, or which cause a gene product to be inappropriately regulated, are often lumped together under the term ***gain-of-function mutant***. While this term is in general use, it seems to us preferable to use the term ***heterochronic mutant*** to describe mutants that cause genes to be expressed at the wrong times. We can then reserve the term *gain-of-function mutants* for mutants that, for example, remove the regulatory element from some signal transduction protein and thus lock that protein in the ON state, or which combine components from more than one gene to create a product with novel specificities for binding partners or active sites in the cell.

1.1.3 DNA-level terminology

The ultimate classification scheme for mutations requires the DNA sequence of the wildtype and mutant alleles themselves. The basic classes of mutations at the DNA level are as follows.

- ***Base pair substitution mutants***: these mutants are a result of the change of one base in the sequence to another. Changes such as A–T to G–C or C–G to T–A that replace a like nucleotide (purine or pyrimidine) on each strand are referred to as *transitions*. Changes such as A–T to C–G, in which a purine on one strand is replaced with a pyrimidine on the same strand (or vice versa), are referred to as *transversions*.

- ***Missense mutants***: these mutants are a class of base pair substitution mutants that, by changing the sequence of a given codon, direct the incorporation of an amino acid different from the one specified at the same position in the wildtype allele.

- ***Nonsense mutants***: these mutants are a class of base pair substitution mutants that alter a given codon to create one of the three "stop" codons UAA, UAG, and UGA. Geneticists old enough to remember the moon landing may sometimes refer to these mutants as *ochre* (UAA), *amber* (UAG), and *opal* (UGA).

- ***Silent substitutions***: these are either mutants in coding that do not change the amino acid directed by that codon, as is often true for *third base substitutions*, or changes in the non-coding region that do not affect gene expression. This class of mutants has taken on real value in providing markers [such as restriction fragment length polymorphisms (*RFLPs*) and other single nucleotide polymorphisms (*SNPs*)] for human genetic mapping. (We might also note that even mutations that do direct amino acid changes might be phenotypically silent if the substitution involves a chemically similar amino acid.)

- ***Base pair insertions or deletions***: the name says it all. Larger deletions are also sometimes referred to as deficiencies.

- *Frameshift mutants*: these result from the insertion or deletion of base pairs from within the coding sequence that alter the reading frame [for a detailed consideration of the genetics of frameshift mutations, see chapter 4 (section 4.2.1)].

We should also note that a series of terms used to describe molecular events have been borrowed from the terminology used to describe chromosome rearrangement. The meanings of most of these terms, such as *inversion*, *duplication*, and *deletion* (or *deficiency*) are self-evident. The term *translocation* refers to a breakage and rejoining event involving sequences on non-homologous chromosomes.

1.2 Dominance and recessivity

We use the term dominant to mean that if **Aa** individuals or cells are phenotypically similar or identical to **AA** cells or individuals, while the **aa** genotype confers a different phenotype, then **A** is the dominant allele and **a** is the recessive allele. If **Aa** individuals or cells are phenotypically intermediate between their **AA** and **aa** counterparts, then **A** and **a** are said to be semidominant. These are seemingly straightforward terms that all of us learned in high school biology. Unfortunately, their usage is often rather careless. One needs to think about the situation in which the term is being applied.

To a certain extent, the terms dominant and recessive are simply matters of perspective. Suppose we look at a loss of function mutant in a human that encodes an essential metabolic enzyme. For reasons that we will detail below, heterozygosity for simple loss-of-function mutants in that enzyme is likely to have little effect on the metabolism in most cell types. Thus, heterozygotes for this mutation are likely to be normal, and we would classify this mutation, at least for purposes of constructing a human pedigree, as fully recessive. But if we refocused our interest only on the amount of active enzyme produced by a given cell type, then our loss-of-function mutant might be codominant. Indeed, in the absence of cellular controls that limit the level of enzyme production, one might expect that virtually all mutants could be shown to be codominant at the molecular level. Thus our terminology only has meaning when put into proper perspective.

A dramatic example of the importance of "perspective" can be seen in the human hereditary cancer disorder retinoblastoma. The inherited form of retinoblastoma is inherited as a simple autosomal dominant (Rb−) and the disorder behaves in a pedigree as a dominant mutation should. However, in affected individuals only ten or so cells in the retina of each eye form tumors in Rb+/Rb− individuals. These cells only become tumorous because they have, by subsequent somatic loss or mutation, lost the normal Rb+ allele and unmasked the inherited Rb− allele. *At the cellular level the Rb− mutation is fully recessive.* The Rb− mutation only induces tumor formation when the wildtype allele is removed! The requisite somatic mutation/loss events are rare. But because there are hundreds of millions of retinal cells in each eye, the somatic mutation events required to unmask the Rb− allele become a virtual certainty to occur somewhere. So are Rb− mutants dominant or recessive? The answer depends on your perspective: in pedigrees they are dominant; in cells they are recessive.

Because it is exactly one's perspective that matters here, we will begin our discussion of dominance and recessivity at the level of the individual cell.

1.2.1   *Dominance and recessivity at the level of the cell*

The cellular meaning of dominance

The critical point is that dominance will result whenever: (i) a single copy of the wildtype gene is insufficient; (ii) the product of the mutant gene is poisonous to the process; or (iii) the mutation causes the gene to be misexpressed in a fashion that creates a phenotype. We have tried to make the point that all classes of mutants can, in the right context, be dominant. Neomorphs and antimorphs are dominant, almost by definition. Hypermorphs could conceivably be dominant, but one could easily imagine cases where they are recessive. Nullomorphs or hypomorphs in genes that encode structural or other dosage-sensitive proteins could also be considered dominant.

So how can you classify a dominant mutation you have just recovered?

- The first and best question is to determine if it is mimicked by a deficiency. If *Df/M* looks like *M/M*, then your mutant is most likely a nullomorph. If *Df/M* is more severe in phenotype *M/M*, then what you have is a hypomorph. If a deficiency does not mimic your mutant, then consider the remaining three cases.
- If a duplication of your gene, i.e. +/+/+, looks like *M/+*, you have a hypermorph.[2]
- If your mutant cannot be mimicked by a deficiency or by a duplication, then the odds are that you have an antimorph or a neomorph. Antimorphs can often be partially suppressed by adding extra copies of the wildtype gene; neomorphs cannot.

The cellular meaning of recessivity

There is a simple reason that most mutants are recessive. Most genes encode enzymes, and for most of the enzymatic reactions that take place in a cell the limiting factor affecting the amount of product (or even the rate of product accumulation) is substrate concentration. So, twofold reductions in the concentration of enzyme itself often have minimal effects on the rate of product formation, and thus on the phenotype. For that reason it is often of little consequence to a cell whether there are one or two functional copies of a gene. (As noted above, mutants in genes that produce structural proteins required in stochastic amounts are not expected to be recessive.)

But surely there are some cases where even a 50% reduction in enzyme concentration might be sufficient to produce a phenotype? Indeed, one could imagine that heterozygosity for a null mutant in an enzyme-encoding gene might place the cell (or individual) near some threshold of a function, making it more susceptible to the effects of genetic or environmental background. Many examples are known of mutants, known as **enhancers**, which allow a normally recessive

---

2   If you find a true hypermorph, please e-mail and tell me about it. Better yet, just send it to me.

mutant at another gene to exert a strong phenotypic effect, even when heterozygous with a wildtype allele at that gene. Consider two genes **A** and **B**. You have recovered loss-of-function alleles of gene **A**, denoted **a**, in a background homozygous for the **b** allele of gene **B**. In this background the **a** allele is recessive. However, you now move the **A** and **a** alleles into a genetic background carrying the **B** allele, only to discover that the **a** allele is now dominant. In other words:

Aa bb wildtype

aa bb mutant

Aa Bb mutant   *a became dominant*

AA Bb wildtype

[Such interactions can reveal a great deal about the functions of the products of gene **A** and gene **B**. We will return to a consideration of this type of interaction in chapter 3 (section 3.4).]

Similarly, there are examples in which recessive mutants can exert strong phenotypes at extremes of the environment, such as high or low temperatures, even when heterozygous with wildtype alleles. There is a mutant in Drosophila that produces small (miniature) wings. Under most conditions, this mutant is fully recessive, but raise a *mutant*/+ heterozygote fly at 16–18°C, instead of the usual 24°C temperature, and the flies have very small wings indeed! The reduction of protein products in mutant heterozygotes fell near a critical threshold. Once an environmental stressor (cold temperature) was applied, the effect of the reduction in gene copy number was revealed. The sickle cell mutation in humans, which becomes dominant at high altitudes, is another example of this type of genotype–environment interaction.

Such cases are not common. The vast majority of loss-of-function mutants are phenotypically recessive. One of the virtues of diploidy is that we apparently can tolerate such 50% reductions in protein production for many, if not most, of our genes. Still, one presumes that many 50% reductions in the normal gene dosage, as would occur in individuals heterozygous for large deficiencies, might be deleterious. Indeed, in a careful study of the effects of duplication and deletion in the Drosophila genome, Lindsley et al. (1972) estimated that flies could not usually tolerate heterozygosity for autosomal deficiencies that were much larger than 1% of the genome (~150–200) genes.

1.2.2   Difficulties in applying the terms "dominant" and "recessive" to sex-linked mutants

Many organisms tolerate hemizygosity for the genes residing on the X chromosome (e.g. sex-linked genes in XY male flies, worms, mice, or humans). Their ability to do so reflects the capacity of each of these organisms to perform one of a set of processes referred to as **dosage compensation**. In the case of flies and worms, the dosage compensation mechanisms act to increase the activity of X chromosomal genes in XY or XO males, respectively, to match that observed in XX females or hermaphrodites. In mammals, the cells of XX individuals (females) shut down one X at random in each somatic cell so that all individuals, both male and female, are effectively hemizygous.

XY hemizygosity

all in rehme: mosaic. The one able to "see" colors give the vision to the others

The fact that mammalian females are mosaic, with each cell containing one active and one inactive X chromosome, raises a serious question as to what we mean when we refer to an X-linked mutation, such as *colorblindness*, being recessive. In the case of colorblindness we mean that, even though each retina of a heterozygote is a mosaic of cells that possess or lack the light-receptor protein produced by this gene, the cells that do produce the light receptor are sufficient to confer color vision on the organism. A similar argument can be made with heterozygotes for mutants in the X-linked clotting factor gene that result in hemophilia. In heterozygous females, having half the blood cells able to produce clotting factor is apparently enough. In each of these cases, our only meaningful use of the terms dominant and recessive is at the organismal level.

1.3   The genetic utility of dominant and recessive mutants

Recessive loss-of-function mutants can be a geneticist's best friend. Understanding the consequences of the loss of activity of a protein can give insights into the function of that protein. That said, the value of dominant mutations in genetic analysis cannot be overstated. Having either antimorphic or neomorphic alleles of a gene allows one to create null alleles of that gene (by screening for revertants of the dominant alleles). The easiest way to revert a poisonous antimorphic mutation, or a differentially expressed neomorphic mutation, is to prevent that allele from encoding **any** protein (by creating a null allele). If you were initially blessed with either an antimorphic or a neomorphic allele of your gene, this is probably the best approach to isolating a null allele of your gene. Dominants also lend themselves to screens for mutants that enhance or suppress the dominant allele, and such screens are good ways to isolate new genes whose protein products participate in the pathway or process under study.

Summary

Our focus on terminology in this chapter was designed less as a didactic lesson in "naming things," than as a lesson in the fact that we define things in terms of how we understand them. Muller understood mutants in terms of phenotypes, both on their own and in combination with other mutants. Thus, his terminology is framed in terms of what mutants "do." Molecular biologists understand things in terms of lesions in DNA, and hence have a rather different nomenclature. Each of these terminologies is valid, but one needs to be careful in choosing terms that match one's understanding of a given mutant. This is especially true with respect to the terms "dominant" and "recessive." These terms have very different means from the perspectives of a cell or a pedigree. To avoid confusion, make sure that the term you use describes an observation or result that matches your nomenclature system, and define the perspective you are using.

Gallery of model organisms

1.1 Our favorite organism: *Drosophila melanogaster*

D. melanogaster was the organism used by Calvin Bridges to verify the Chromosome Theory of Heredity (Bridges 1914, 1916). Since then it has remained as a central player in genetic research. Although there are other species in the Drosophila group (Ashburner 1989), when we use the terms Drosophila, fruit fly, or simply flies in this narrative we will always mean *D. melanogaster*.

Basic culture techniques

Drosophilae are easily raised on a cornmeal–molasses medium (Ashburner 1989). They are not harmful to humans, nor do they carry parasites that are harmful to people. Under ideal conditions an egg laid by a female fly will develop into an adult fly in 9 days. This includes three larval instars and several days of pupation before hatching. Females are competent to mate within 8–12 hours after hatching. Unfortunately, they also store sperm from each mating they experience and will use that sperm throughout their lives. Thus to set up controlled matings one needs to collect virgin females (i.e. females who have had no previous opportunities to mate with males of a different genotype). This can be done easily by clearing bottles of hatching flies in the morning and then collecting recently hatched females in the late afternoon. Once mated a single female can produce several hundred eggs in a matter of 1–2 weeks. A basic introduction to working with Drosophila can be found at: http://www.ceolas.org/fly/intro.html.

Useful guides and manuals

There are literally tens of thousands of genetically marked strains of Drosophila. The best guides to the available mutants are the so-called redbooks: Lindsley and Grell (1968) and Lindsley and Zimm (1992). One can also get constantly updated information on mutants, chromosomes, and strains from several websites: Flybase (http://flybase.bio.

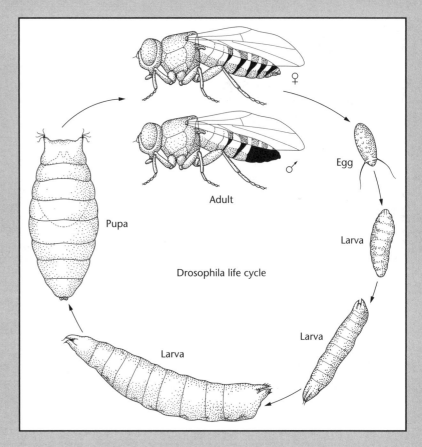

Fly life cycle

indiana.edu/) and from the website for the Bloomington Drosophila Stock Center at Indiana University: http://flystocks.bio.indiana.edu/.

Nomenclature

Fly nomenclature is decidedly complex. The rules are, however, clearly spelled out at http://flystocks.bio.indiana.edu/. For our purposes, and over-simplifying greatly, we need only note that gene names and symbols are always printed in *italics*. For recessive mutants, the name and the symbol are fully in lowercase (e.g. *white*, denoted *w*, or *bithorax*, denoted *bx*). In the case of dominant mutants, the first letter of the name or symbol is capitalized (e.g. *Bar*, denoted *B*, or *Abnormal X segregation*, denoted *Axs*). A wildtype or "normal" allele is denoted either simply by a + symbol in a heterozygote (e.g. *nod*/+) or by a superscript following the gene symbol (e.g. the wildtype of *nod* could be denoted as *nod*$^+$). Protein products that are named for the gene that produces them are also symbolized by the gene symbol, but this symbol is all in capital letters. When the full gene name is used for the protein, rather than the gene symbol, only the first letter of the name is capitalized. For example, the protein product of the *nod* gene could be correctly denoted as the *nod* protein or Nod.

When genes on more than one chromosome are represented in a genotype, the markers on each chromosome are separated by semi-colons in the order X;Y;2;3;4. Gene symbols are separated by a space. Homologs are separated by a forward slash (/). Homozygous chromosomes are defined only once, and + implies +/+. Thus a female heterozygous for *y* (an X chromosome gene) and for *cn* (a 2nd chromosome gene) would be denoted *y/+;cn/+*, while females homozygous at both loci would be denoted *y;cn*.

Chromosome biology

D. melanogaster has four chromosome pairs, three autosomes and one pair of sex chromosomes. Although males are XY and females are XX, sex is essentially determined by the number of X chromosomes and not by the presence or absence of the Y chromosome. The Y chromosome is only required for male fertility. Thus XO animals are sterile males and XXY individuals are fertile females. The well-known polytene chromosomes are found most easily in the salivary glands of third instar larva. Only the euchromatic regions of the chromosomes (approximately the distal two-thirds of each arm) amplify in the polytene chromosomes. The heterochromatic region appears at best as dense threads at the base of each chromosome. Mitotic chromosomes are easily found in early embryos and in dividing cells of the 3rd instar larval brain. Techniques for visualizing meiotic chromosomes in both sexes are now well developed. Methods for such an analysis of chromosomes in flies are copiously provided by Sullivan et al. (2000).

A few genetic idiosyncrasies

Perhaps the most critical things for readers of this book to remember are the following:
1 There is **no** recombination during meiosis in Drosophila males.
2 There is no recombination on the small 4th chromosome in Drosophila female meiosis.
3 The X, 2nd, and 3rd chromosomes do recombine during female meiosis, and on average there is slightly more than one recombination event per chromosome arm.
4 The Y chromosome does not determine sex, but does carry genes essential for male fertility.
5 Flies that carry only one 4th chromosome do survive, but they are sterile. They also exhibit a strong *Minute* phenotype (thin bristles, delayed emergence). Flies with three copies of chromosome 4 are fully viable and fertile.

6   XO flies are sterile males; XXY flies are fertile females; XXX females survive but are sterile.

7   Flies carrying either one or three copies of the major autosomes are not viable. (The only exception to this rule is full triploids, which do survive.)

1.2   Our second favorite organism: *Saccharomyces cerevisiae*

Because the baker's, or budding, yeast is a eukaryote that can be handled with the same facility as *Escherichia coli*, this organism has been the workhorse for much of the development of modern eukaryotic genetics. The ease of culture and ability to grow as either a stable haploid or diploid make this organism ideal for mutant screens. Indeed, the ability to grow this yeast as a haploid at various temperatures, and to replicate plate colonies, makes it straightforward to isolate both loss-of-function and conditional mutations at a large number of loci. Complementation tests are easily done by mating haploids to diploids, and meiosis can be studied by inducing the diploid culture to undergo meiosis (sporulation). Sporulation results in the production of a walled ascus containing four spores: the four products of meiosis. As we will see in chapter 7 (section 7.4.3), the ability to recover all four products of meiosis greatly facilitates the analysis of the meiotic process.

Basic culture techniques

Yeast can be grown in a standard broth medium and on Petri dishes. The haploids exist as two genetically determined mating types (a and alpha). Haploids of opposite mating type can be mated to produce stable a/alpha diploids that can be sporulated by nitrogen starvation.

Useful guides and manuals

The best guide to the genetics and biology of this organism may be found at a website maintained by Fred Sherman at the University of Rochester: http://dbb.urmc.rochester.edu/labs/Sherman_f/yeast/1.html. Basic information on yeast culture and genetics can be found in the *Methods in Enzymology* volume edited by Guthrie and Fink (1991). The central repository for current information regarding yeast genes and gene sequences is the Saccharomyces Gene Database (SGD): http://genome-www.stanford.edu/Saccharomyces/yeast_info.html. SGD maintains a virtual library of information related to yeast and yeast culture: http://genome-www.stanford.edu/Saccharomyces/VL-yeast.html.

Nomenclature

Genes are normally named by a three-letter code describing some aspect of their isolation, their phenotype, their required function, or the protein

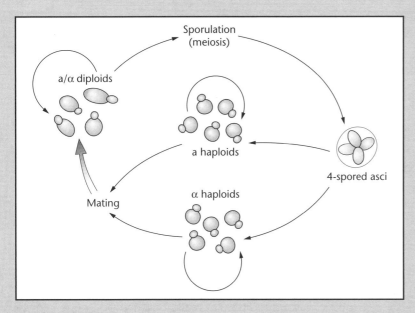

Yeast life cycle

they produce. Thus a gene defined by arginine auxotroph mutant would be symbolized as ARG, a gene encoding actin as ACT, and a gene required for the cell division cycle as CDC. In cases where two or more different genes have the same letter symbol, they are distinguished by numbers, such as URA1 and URA2. Alleles of a given gene are indicated by a number following the gene system and separated from the symbol by a hyphen, for example *cdc2-1* and *cdc2-2* are different alleles of the CDC2 gene. Dominant alleles are symbolized using uppercase italics (e.g. *CDC2*). Recessive alleles are denoted by lowercase italics (e.g. *cdc2-1*). Wildtype genes are designated with a superscript "plus" (e.g. *ACT2⁺*).

Chromosomes

The yeast genome is comprised of 16 well-characterized chromosomes. Unfortunately, the chromosomes are too small for standard metaphase chromosome analysis. However, bivalent chromosomes can be visualized during meiosis.

1.3 Our third favorite organism: *Caenorhabditis elegans*

The worm *C. elegans* is one of the most important model organisms in use today. This tiny worm lives on *E. coli* (of all things) and is most easily cultured on Petri dishes. *C. elegans* has two sexes, hermaphrodites (XX)

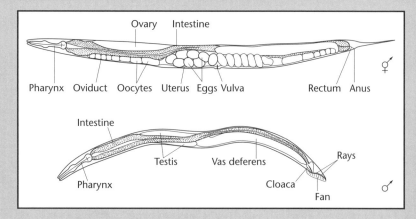

C. elegans males and hermaphrodites

and males (XO). Hermaphrodites can self-fertilize or mate with males. When they mate with males, half the progeny are males and the other half develop as hermaphrodites. However, one hermaphrodite cannot mate with another hermaphrodite. The existence of hermaphrodites is arguably the most useful genetic feature of the system, considering that a zygote resulting from a mutagenized game will develop as a heterozygote. If that heterozygote is a hermaphrodite, it will produce homozygotes in the next generation. Thus one can go from mutagenesis to homozygotes for mutagenized chromosomes with just one intervening generation.

Basic culture techniques

Development from an egg to a fertile adult hermaphrodite or male takes only 3 days, and a hermaphrodite can produce about 300 offspring using self-fertilization alone. In the presence of males, even greater numbers of progeny are easily obtained. When hermaphrodites mate with males, 50% of the progeny will be males and 50% will be hermaphrodites. The organism is transparent and contains an invariant number of somatic cells (959). The lineage relationships that produced the adult animals are fully described. Best yet, stocks of any given genotype can simply be stored frozen, avoiding the time-consuming process of stock keeping.

Useful guides and manuals

The book *C. elegans II* by Riddle et al. (1997) is available online at http://www.ncbi.nlm.nih.gov/books/bv.fcgi?call=bv.View..ShowSectin-&rid=ce2 and provides an invaluable reference manual for *C. elegans* biology. More detailed information on a given strain or mutant may be obtained by accessing Wormbase at http://www.wormbase.org/. A brief review of *C. elegans* genetics and nomenclature can be found on Mark Blaxter's website at: http://nema.cap.ed.ac.uk/Caenorhabditis/ C_elegans.html#top.

Nomenclature

As in the case of yeast, genes are normally named by a three-letter code describing some aspect of their isolation, their phenotype, their required function, or the protein they produce, followed by a dash and then a number. So the fifth gene defined by an *uncoordinated movement* mutant (*unc*) would be named *unc-5*. Gene names and symbols are *italicized*. Alleles are denoted by single or double letters, identifying the laboratory that isolated the mutant, followed by a number. So an *unc-5* allele isolated in Kansas City might be called *unc-5(kc001)*. Wildtype alleles are denoted by a + symbol, and the protein product of a given gene is symbolized in all capital letters (no italics, i.e. UNC-5).

Chromosomes

C. elegans has five pairs of autosomes and one pair of sex chromosomes. The chromosomes are holokinetic: diffuse centromeres during mitosis, but unikinetic (single centromere) during meiosis (Albertson & Thomson 1993).

1.4   Our new favorite organism: zebrafish

Concomitant with all of the benefits of doing genetics in Drosophila and yeast, there are limitations. While the small size of flies makes it possible for a single researcher to easily maintain hundreds of thousands of Drosophilae, it also eliminates the need for *D. melanogaster* to have systems such as an endoskeleton, or a closed circulatory system. Thus, workers have begun to look for vertebrate systems that might be handled as easily as Drosophila. One such organism is the zebrafish: *Danio rerio*. Zebrafish afford us a number of the same advantages that we enjoy when working with *D. melanogaster*, but that systems such as mouse lack. For instance, zebrafish are easy to maintain, can produce thousands of embryos in a week, and perhaps most importantly, the zebrafish embryo is transparent. This affords us the possibility of studying development unencumbered in embryogenesis. All of these things, in combination with the ever-increasing collection of genetic markers and molecular methods, are quickly making zebrafish one of the most popular vertebrate model organisms.

Basic culture techniques

Zebrafish are easily maintained in tanks ranging from one to 40 L in size, and can be fed many types of food including spirulina flakes, and even

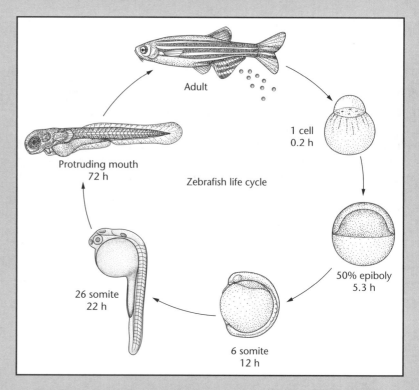

Adult

1 cell
0.2 h

Protruding mouth
72 h

Zebrafish life cycle

50% epiboly
5.3 h

26 somite
22 h

6 somite
12 h

Zebrafish life cycle

fruit flies (Westerfield 2000). This is a very handy way of feeding your fish if you work in the same building as a fly laboratory. Fly geneticists are all too willing to relinquish old bottles of flies, for which they no longer have need. The best food, however, if you want to keep your fish happy, is live brine shrimp. Like graduate students, zebrafish seem to survive better when kept on a regular day–night schedule of 14 hours of light for every 10 hours of dark. This means that if you don't have an entire room to devote to zebrafish stocks, you will have to get lighting for the tanks, and covers to block out ambient light created by your fellow scientists who might be working late.

Useful guides and manuals

One of the most useful guides in working with zebrafish is *The Zebrafish Book* by M. Westerfield (2000). This volume covers everything from general fish care to histological, genetic, and molecular methods. This book is available online through http://zfin.org, a website that also contains information about available mutant and wildtype strains, molecular markers, recent publications, and laboratory contacts.

Nomenclature

Gene names in zebrafish are written lowercase, in italics. In general, gene symbols are three-letter lowercase abbreviations. For example, the gene *albino* is abbreviated as *alb*. Wildtype alleles of genes are given by the three-letter abbreviation, followed by a superscript +. To continue with our example, the wildtype allele of the gene *albino* would be designated *alb*$^+$. Mutant alleles are also given by a three-letter designation, but followed by a superscript code which generally gives an allele number, and a designation of where the mutant was isolated (for a complete list of codes for where genes are isolated, refer to http://zfin.org). For example, the allele *alb*^{b4} is named as such because *b* is the code for Eugene, Oregon, which is where this allele was isolated. Dominant mutant alleles are named with a *d* as the first letter in the superscript. For example, *nbb*^{da15} is a dominant allele of the gene *night blindness b*. Protein products of genes are simply written as the three-letter gene code, not italicized, with the first letter capitalized.

Chromosome biology

Zebrafish have 25 linkage groups, each corresponding to one of the 25 haploid chromosomes. Zebrafish males and females appear to have the same complement of chromosomes, i.e. no sex chromosomes, unlike flies or humans.

1.5   Phage lambda

The discipline of developmental genetics focuses on understanding how cells make choices during development, such as differentiating along a male or a female pathway, becoming a wing or a leg progenitor, etc. For quite some time the best model that geneticists had for understanding the genetic basis of such choices was phage lambda. Lambda is a temperate phage, meaning that following infection it can either replicate and then lyse the bacterium, or it can eschew replication and other lytic functions and integrate itself into the bacterial chromosome in a stable and heritable fashion. This latter state, known as **lysogeny**, allows the virus to remain a quiescent component of the bacterial genome for long periods of time. The integrated copy of the lambda genome is referred to as a **prophage**. Following infection, this choice of fates is indicated by cloudy plaques made up of healthy lysogenic bacteria surrounded by a sea of lysed corpses resulting from lytic infections. The prophage remains repressed until, as a result of a variety of physiological changes or environmental insults, it is **induced** to exit the bacterial genome and enter a lytic cycle.

Thus each time a lambda phage enters a cell it must choose between a lytic and a lysogenic pathway. This then becomes a paradigm for a

Chapter 1

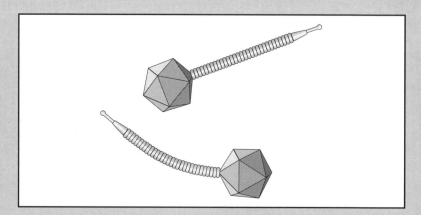

Phage lambda

genetic "switch" involving primarily the products of two genes: *cI*, which encodes a repressor of lytic functions; and *cro*, which encodes a protein that (among other things) blocks expression of *cI*. Mutants in the *cI* gene result in clear plaques because all the infections result in lysis; hence the name of this gene, *clear-I*. Mutants in *cro* favor the establishment of lysogeny. As infection occurs there is a "race" to produce a sufficient abundance of cI protein to shut down lytic functions, including production of cro protein, against producing enough cro protein to block the production of a sufficient amount of cI protein to establish repression. [Readers interested in this fascinating switch are referred to a book entitled *A Genetic Switch* by Mark Ptashne (1987).]

Once established, the lysogenic pathway involves the integration of a circular lambda chromosome into the bacterial host chromosome by site-specific recombination at unique attachment (*att*) sites on the phage and bacterial chromosome. The continued production of cI protein maintains the prophage in a repressed state following integration, and thus allows the propagation of the lysogen. This state is stable until environmental factors lead to a dramatic reduction in the amount of cI repressor protein, and thus facilitate a switch to a lytic mode. For example, exposure of the cell to UV light can induce a repair system that cleaves the repressor, permitting expression of lytic functions including those required to excise the prophage from the chromosome. The circular phage then replicates as a rolling circle, producing long concatomers of phage that are then packaged into waiting capsids.

Packaging of mature phage into phage heads involves cutting at specific sites, referred to as *cos* sites (for cohesive ends), a fashion that generates complementary single-stranded tails at both ends of the molecule. Thus, unlike the case described for phage T4 (below), each capsid contains a full-length DNA molecule with a defined sequence.

Upon infection, the *cos* sites allow the phage DNA molecule to circularize and begin the cycle of lysis and infection again.

Nomenclature

In general, in bacterial and phage genetics the phenotype is described by a three-letter code, with the initial letter in uppercase. A symbol or letter superscript is often added to further describe the defect. Thus, Leu– means a leucine auxotrophy and kan^R denotes kanomycin resistance. Genes are also referred to by three-letter italicized codes, in which the first letter is not capitalized. An additional capital letter suffix is then added to denote the specific gene in question; thus *araA* would refer to arabinose gene A. The "minus" sign is implied here. The lambda nomenclature evolved far ahead of the nomenclature convention. Many lambda genes are denoted by one- or two-letter symbols, often followed by a number to denote the specific gene (e.g. *cI*).

Useful guides and manuals

The best sources of information on lambda are found in *Lambda II*, by Hendrix (1983).

It is impossible to overstate the role that lambda genetics has played in developing our understanding of gene regulation, recombination, and replication. Moreover, derivates of lambda were the original Charon phage workhorses of cloning vectors, and highly engineered derivatives of lambda comprise many of the modern viral cloning vectors. Work on lambda continues to this day. It remains a shining example of the real power of prokaryotic genetics.

1.6   Phage T4

With a 22-min life cycle and the ability to produce 300 progeny phage for each infected bacterium, this bacteriophage is in many ways a geneticist's dream organism. T4 is a double-stranded DNA phage with a genome of 166 kilobases. The total map length is quite high (1,500 map units), and thus recombinants are easily obtained between any two separate mutations.

The most curious feature of the genetics of this organism is that each phage carries a ds DNA molecule that is approximately 1% longer than the full-length chromosome, thus creating a small terminal repeat. The sequences present in this repeat vary from phage to phage because the genome of phage T4 is circularly permutated. Thus, for several different phage the sequences might be:

<u>1 2 3</u>................................24 25 26 <u>1 2 3</u>
  <u>3 4 5</u>..........................24 25 26 1 2 <u>3 4 5</u>
    <u>4 5 6</u>........................24 25 26 1 2 3 <u>4 5 6</u>
            <u>12 13 14</u>..................1 2 3 4 5 6 7 8 9 10 11 <u>12 13 14</u>

The function and generation of these repeats may be understood by considering the mode of replication. Once a single ds DNA molecule enters the cell, the replicated copies of that single DNA molecule can repeatedly recombine, using the terminal repeats, to produce large concatomers, many phage genomes in length:

<u>1 2 3</u>.....................24 25 26 <u>1 2 3</u>
                      X
                      <u>1 2 3</u>...........................24 25 26 <u>1 2 3</u>
<u>1 2 3</u>.....................24 25 26 <u>1 2 3</u> <u>1 2 3</u>..................24 25 26 <u>1 2 3</u>

Once the long concatomer is created, the packaging of phage into waiting capsids, or heads, begins at one end of the concatomer. Each capsid "swallows" slightly more (1%) than a full-length phage genome, thus recreating the terminal repeat. But because packing is processive, each new capsid will carry a phage with a different terminal repeat (see figure below). Thus, the resulting phage burst will include a population of phage carrying different circularly permuted genomes, each with a terminal repeat.

The discovery of circular permutation and terminal repeats resulted from the observation that, following a mixed infection of phage differing at some locus (say A and a), it was possible to obtain progeny phage carrying two different copies of a given gene (when that gene is included in the terminal repeat). Infection with such a heterozygous phage will yield a "mixed burst" of wildtype and mutant progeny.

Students occasionally ask how this peculiar feature of T4 genetics impacts on the *rII* experiments discussed at several places in this book. We must understand that screens for wildtype recombinants will not be affected by the creation of terminal repeats. (Note that the screen for a

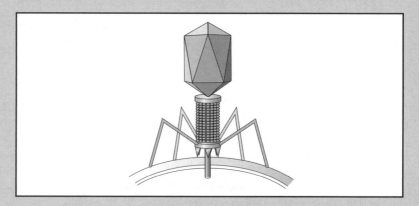

Phage T4

wildtype recombinant requires that progeny phage replicate on the restrictive bacterial host. Thus, even in rare cases where a wildtype recombinant allele is present in a terminal repeat, and is thus heterozygous with a mutant parental allele, once the infection occurs only wildtype bearing progeny phage will contribute to the resulting plaque. After the first cell is lysed only the wildtype phage will be able to continue to infect neighboring cells.) Similarly, complementation studies require the mixed infection of two pure strains. The presence of terminal repeats for the mutant alleles on 1% of each of the two parental input phage will have no effect on these experiments.

Phage T4 continues to be an excellent system in which to study the fundamental processes of DNA metabolism. In addition, the biological processes that allow the assembly of the syringe-like phage have provided key insights into the mechanisms by which large protein complexes are assembled.

1.7   *Arabidopsis thaliana*

In less than a decade this small mustard weed plant has become one of the premier genetic organisms. A combination of a small chromosome number (5), a sequenced genome, and a small genome size (114.5 megabases) make it attractive as a genetic tool. Add to that a short generation time (approximately 6 weeks), the ability to grow under defined conditions, a very high fecundity (up to 10,000 seeds per plant), and the relative ease of propagation, and the attraction becomes irresistible. Efficient transformation systems exist, and there are both a large number of available mutants and several stock centers available.

Mutagenesis in Arabidopsis can be done using chemical (EMS) or physical mutagens (neutrons), but is more often done using transposable elements or random T-DNA integration following exposure of the plant to Agrobacterium. An excellent review of these methodologies may be found in Page and Grossniklaus (2002). Indeed, EMS mutagenesis of seeds works sufficiently well that after mutagenizing seeds, screens of 2,000–3,000 plants in the next generation usually yield two or three new mutants at most loci.

The facility of genetic analysis in this organism allows screens for enhancer and suppressor mutants as well as for sophisticated epistasis analysis. Until recently, the analysis of meiotic chromosome behavior had been hindered by the lack of useful protocols for meiotic cytology. Thanks largely to the work of Gareth Jones and his collaborators (Armstrong et al. 2001; Sanchez-Moran et al. 2001), this difficulty has been fully overcome.

Useful guides and manuals

The best sources for printed information are *Arabidopsis*, edited by E. M. Meyerowitz and C. R. Somerville (1994) and *Arabidopsis: A Laboratory*

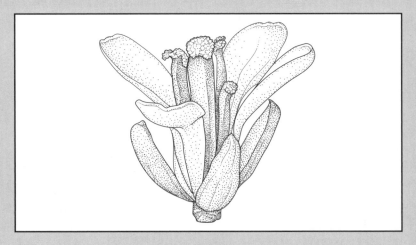

A. thaliana

Manual, by D. Weigel and J. Glazebrook (2002). Web-based information can be found at the Arabidopsis Information Resource at CSHL: http://stein.cshl.org/atir/biology/genome/ and at the Arabidopsis Inform-ation Resource at Stanford: http://www.arabidopsis.org/home.html.

Nomenclature

Wildtype genes are denoted by three italic letters, followed by a number to differentiate genes with a similar symbol (i.e. *ARB1* or *ARB2*). Mutants are denoted by lowercase italics (*arb1* or *arb2*). Different alleles at the same locus are differentiated by a hyphen and a number (i.e. *arb3-3* would be the third allele of the *arb3* gene). Some workers denote domin-ant alleles by adding a *D* to the end of the symbol. Proteins are denoted using the wildtype gene symbol entirely in uppercase, no italics (i.e. ARB3). Phenotypes are usually denoted by fully writing out the gene name (not italic, but capitalizing the first letter), followed by a superscript (+ to denote wildtype and – to denote mutant). A guide to mutant genes and their nomenclature can be found at http://mutant.lse.okstate.edu/genepage/genepage.html.

1.8   *Mus musculus* (the mouse)

> "Let us never forget that it all started with a mouse"
> Walt Disney

In many ways, early genetics had its beginnings in the work of so-called mouse fanciers who helped to extend the findings of Mendel. The mouse has continued to be a pre-eminent system for genetics, and indeed taking the problem to the level of the mouse is a primary goal of many model organism studies. Some years ago an administrator at the National Institute of Health was overheard to remark, "say what you want about all of these model organisms, they just aren't mice." Like it or not, the mouse has become the premiere model organism for human biology, both because its mammalian biology closely approximates that of human beings, and because of the facility for genetic manipulation that is possible in this organism. The mouse system benefits greatly from a relatively short generation time (~9 weeks), the availability of inbred strains, and the relatively low cost of rearing. [Having written that sentence, we can hear every mouse geneticist that we have ever known screaming about the high costs of raising mice. True, they are more expensive to keep than flies, worms, or zebrafish. But they are cheaper than rats or hamsters, and incomparably less expensive to rear than larger mammals (e.g. blue whales).]

There are hundreds of naturally occurring mouse mutant strains. But the relative ease of targeted mutagenesis in the mouse has led to the availability of thousands of mutant lines. Because mouse embryos can be frozen and kept indefinitely, as can sperm and ova, keeping huge numbers of variant lines becomes technically feasible. The strains are maintained primarily at the Jackson Laboratories (www.jax.org) in the US and at the Harwell Laboratories in the UK. There is an excellent "Bible" for doing mouse genetics. It is *Genetic Variants and Strains of the Laboratory Mouse*, edited by M. F. Lyon et al. (1995). The rules for nomenclature are spelled out on the Jackson Laboratory website (http://www.informatics. jax.org/mgihome/nomen/).

Mutagenesis can be accomplished using either chemical or physical mutagens, but most commonly these days by targeted gene disruptions. A guide to mutagenesis in the mouse can be found in Justice (2000), and more molecular approaches are described in Stanford et al. (2001) and Copeland et al. (2001). Mouse geneticists continuously claim to be on the verge of possessing respectable balancer chromosomes [cf. Yu and Bradley (2001)], but by Drosophila standards there is quite a long way to go in this effort.

M. musculus

Mice have 40 chromosomes, 19 pairs of autosomes, and one pair of sex chromosomes. The sex determination pathway is similar, if not identical, to that of humans, and thus sex is determined by the presence or absence of a single gene on the Y chromosome. It can be difficult to distinguish specific chromosomes with simple (e.g. Giemsa) staining. This difficulty is marginalized by the availability of fluorescent hybridization "cocktails" that uniquely map each chromosome pair by FISH analysis. Male meiosis has long been amenable to detailed cytological analysis. However, in recent years the work of Patricia Hunt and her collaborators has provided elegant (and often artistic) inroads into the cytology of female meiosis as well (Hunt & LeMaire-Adkins 1998).

However, the real value of the mouse lies in the biological similarity of the mouse to the only organism that most of us *really* care about, namely ourselves. Despite a few rather spectacular failures, most mutants made in the mouse model are similar to mutants in humans in terms of a disease phenotype. And one can begin drug studies in the mouse that might often be meaningless in flies or worms. One can also do genetic studies in mice that are either not ethically allowable or technically feasible in humans. So perhaps Walt Disney only got it half right, it all probably ends with a mouse too.

Mutant hunts[1]

"Before beginning a Hunt, it is wise to ask someone what you are looking for before you begin looking for it."

A. A. Milne

The very essence of "doing genetics" is the isolation of mutants. Mutants allow us to look into a genome with 10,000–50,000 genes and identify those genes whose protein products participate in a given biological process. To paraphrase Leland Hartwell, who won the Nobel Prize on the day that we wrote these lines, the ability to identify the "right genes" by the "right mutants" is the *surgical power of genetics*. To understand the significance of this statement, we must understand that a given biological process may require hundreds of different proteins acting at various times and in various tissues during development. Not all of these proteins may be specific to that process; many may function in other processes as well. If we tried to identify the critical players in a given process only by asking for proteins that were abundantly and specifically expressed in a given tissue or at a given time, we would miss many critical players in the process. However, our ability to detect a gene's function by genetic analysis does not require that the protein product be either abundant or tissue specific. Nor do we require that it be induced or repressed by some stimulus in order to identify the gene.

As long as knocking out a gene impairs a product (and thus a process), we can obtain mutants in that gene. Because so many mutants are plieotrophic (i.e. they affect more than one function), mutants can also unexpectedly tie together different biological processes or pathways. The learning mutant that also affects female fertility or eye development may do so because it defines a signal transduction pathway that is common to both processes. Combinations of mutants can be used to elucidate regulatory pathways (via epistasis analysis) that would be difficult, if not impossible, to solve in some other, more molecular, fashion. Thus the first, and most critical, step in understanding any biological process is the isolation of new mutations.

People search for new mutations for four very different reasons. Not surprisingly, each motivation poses different problems and requires a rather different approach. For that reason, we begin with the following question.

---

[1]   And other rainy day diversions.

2.1   Why look for new mutants?

2.1.1   *Reason 1: To identify genes required for a specific biological process*

The first reason to isolate new mutations is to identify the genes required for a specific biological process. Flight, for example, has always fascinated biologists. Anatomists might be concerned with the structure of a hawk's wings. Physiologists may want to examine the mechanics of muscle contraction and wing motion. Somewhere out there, biochemists may be boiling down the wings, pouring the resulting soup over a column, and trying to reconstitute flight in the cold room. But the geneticist may wonder, "Can I find mutant hawks that fly funny, or better yet, don't fly at all? And if I could get such mutants, what might they tell us about the genes, and the corresponding protein products, that are required for flight?"

Thus, your first motivation for isolating mutants might be that you are interested in some biological process, such as flight, meiosis, or the cell cycle, and wish to identify the genes whose products are essential for that process. In this case, you would probably want to identify as many important genes as possible. Your major concern will, in fact, be to determine how many mutants need to be recovered to be assured that most, if not all, of the genes whose products are required for this process are identified. A major component of answering this last question is determining how many different genes are defined by your mutations (e.g. your collection of 20 new mutants might define 20 new genes, or 20 new alleles of a single gene). At this point in your study you are usually not concerned with obtaining specific alleles of a given gene.

The first concerted attempt to acquire a collection of mutations that affected a defined biological process was the efforts of Larry Sandler and Dan Lindsley and their collaborators to isolate mutants that impaired meiosis in Drosophila (Sandler et al. 1968). These workers set out to identify a collection of mutants defective in a specific biological process (meiosis) in Drosophila by screening through natural populations of Drosophila, found near the vineyards in Italy, for mutants that increased the levels of meiotic chromosome missegregation. As they did so, two of Sandler's graduate students were performing a similar screen using chemically mutagenized strains of Drosophila (Baker & Carpenter 1972). These screens identified a large set of new mutants whose subsequent analysis would be critical to increasing our understanding of the meiotic process (Hawley 1993). These efforts would be followed by the seminal screens of Hartwell for cell division cycle mutants in yeast (Hartwell et al. 1970, 1973; Hartwell 1991) and Nusslein-Volhard and her collaborators for segmentation-defective mutants in Drosophila (Nusslein-Volhard et al. 1984). Each of these screens recovered a large number of mutants whose analysis provided critical insights into the process being studied.

The screen by Nusslein-Volhard and collaborators is particularly instructive. Their goal was to identify genes required for early embryonic development in Drosophila. These authors began by searching for a specific class of mutants

Box 2.1   A screen for embryonic lethal mutations in Drosophila

The diagram below describes the "mating scheme" used by Nusslein-Volhard et al. (1984) to collect recessive lethal mutations. They began by treating males homozygous for the mutants *cn* (*cinnabar*) and *bw* (*brown*) with the mutagen EMS. When doubly homozygous the *bw* and *cn* mutations produce a white-eye phenotype. These treated males were then crossed to *DTS91/CyO* females. The code *DTS91* denotes a dominant temperature-sensitive mutant that kills developing (but not adult) flies at 29°C. The code *CyO* denotes a balancer chromosome that suppresses recombination in females:

P   *DTS91/CyO* females × *cn bw* males
(treated with EMS)
↓
F1   *DTS91/CyO* females × (*cn bw*)*/*CyO* males
(one male per vial)

Set up 10,000 of these matings, with one male and several females in each vial.
Raise the progeny at the restrictive temperature (29°C) for the first 4 days.

This will kill all the progeny that received the *DTS91* chromosome. The code (*cn bw*)* denotes a mutagenized chromosome.
↓
F2   (*cn bw*)*/*CyO* females   ×   (*cn bw*)*/*CyO* males

Mate brothers and sisters arising from each F2 cross.
↓
F3   (*cn bw*)*/(*cn bw*)* and (*cn bw*)/*CyO*

In vials where the (*cn bw*)* chromosome carries a new recessive lethal mutation, no *cn bw*/*cn bw* progeny will be produced. Thus, no white-eyed progeny will be seen. However, (*cn bw*)*/*CyO* progeny of both sexes will survive and can be used to create a **balanced stock**. Because the *CyO*/*CyO* progeny die, the only progeny produced by successive brother–sister matings will be (*cn bw*)*/*CyO*. Nusslein-Volhard and collaborators sorted through several thousand lethal-bearing stocks to find those in which 25% of the embryos (the lethal homozygotes) showed unusual morphologies.

(those that confer recessive lethality) and then devised a set of rapid sub-screens to identify the mutations of interest. The screen is discussed in detail in box 2.1. Out of 10,000 mutagenized chromosomes tested, the authors recovered 4,217 chromosomes carrying at least one new lethal mutation. Each of these lines was then tested to determine the time of death. Of these, approximately half (2,843) caused death in early embryos. Of these 2,843 lines, 321 carried mutations that were both embryonic lethal and exhibited clearly abnormal morphology among the homozygous embryos. Nusslein-Volhard and collaborators would go on to show that these 321 mutants defined some 60 genes whose products were critical for early events in Drosophila development. (Structurally aberrant chromosomes, known as **balancer chromosomes**, were critical to the success of this screen, and of most mutant screens in flies. The nature of balancers is discussed in box 2.2. A variant of this scheme, designed to isolate lethal mutants on the X chromosome, is presented in box 2.3.)

Lee Hartwell began his search for mutants in genes that controlled the yeast cycle by casting an equally wide net for any mutant that was temperature sensitive for growth. He then screened a collection of 1,500 *ts* mutants, and screened through cultures of each mutant, raised to the restrictive temperature, looking for the "rather uniform and sometimes unusual cellular and nuclear morphologies

Box 2.2   The balancer chromosome

For Drosophila genetics the single most critical tool for the analysis of mutants is the balancer chromosome. A good balancer consists of three elements:

1   A set of overlapping inversions that suppress both the occurrence and recovery of crossovers.
2   An easily recognized dominant mutation that allows the balancer to be easily followed in crosses.
3   A recessive lethal or sterile that prevents balancer homozygotes from surviving. [Note that X chromosomal balancers in Drosophila do not usually carry a recessive lethal (for obvious reasons), rather some of them carry a recessive female sterile mutation.]

Inversions suppress crossingover for a variety of reasons. Small paracentric inversions (those that do not include the centromere) actually suppress the occurrence of recombination, especially in the regions surrounding their breakpoints (Novitski & Braver 1954; Theurkauf & Hawley 1992). Larger paracentric inversions suppress crossingover by preventing the recovery of crossover chromatids (Sturtevant & Beadle 1936). As shown in figure B2.1, the crossover event within the inversion cre-ates a dicentric bridge and an acentric fragment, both of which are excluded from the meiotic pro-duct that is made available to the sperm for fertil-ization. Exchange within pericentric inversions creates large duplications and deficiencies (figure B2.2) that are virtually always lethal to the zygote.

Most good balancers combine both pericentric and paracentric inversions, and most balancers have enough breakpoints that they do indeed suppress the actual occurrence of exchange. But some autosomal balancers are less effective at blocking crossovers than others, especially near the ends of the arms. So be careful to choose a good balancer.

All balancers carry a strong, easily recognized dominant mutation, or claim to. But some "easily recognized dominants" are easier to recognize than others. All autosomal dominants also carry at least one recessive lethal mutation to prevent homozygosity. Readers interested in choosing a balancer for a given purpose are referred to Lindsley and Zimm (1992), or to Flybase. There are efforts underway to build balancer chromosomes in the mouse (Zheng et al. 1999; Yu & Bradley 2001) and in the worm (Edgley et al. 1995).

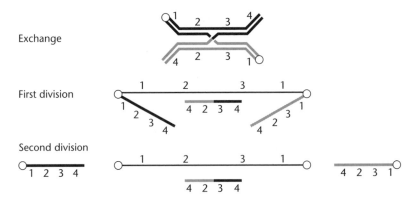

Exchange

First division

Second division

Figure B2.1   A crossover event within the inversion creates a dicentric bridge and an acentric fragment. The diagram is schematic, for instead of an inversion loop formed when the chromosomes are paired throughout their length, synapsis is indicated only for the inverted segment. This makes the consequences of the SCO, formation of a dicentric bridge and acentric fragment readily apparent

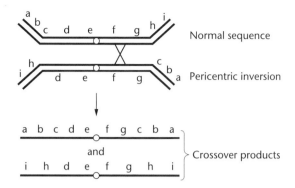

Figure B2.2 Exchange within pericentric inversions creates large duplications and deficiencies that are virtually always lethal to the zygote

that the mutant cells assume after an interval of growth at the restrictive temperature" (Hartwell et al. 1973). The assumption was that those colonies bearing a *ts* mutant in a common housekeeping function would simply die randomly in the cell cycle, but mutants in genes required for progression through the cell cycle would simply arrest at that control point. He recovered 148 such cultures that defined 32 different genes.

As evidenced by the success of these, and subsequent, large screens, the right collection of mutations can provide a strong basis for the analysis of a given biological process. A brief list of references for mutant hunting in a large number of organisms is provided at the end of the chapter in box 2.5. One just needs to be able to isolate the mutants, and then know what to do with those mutants once they are obtained. RSH's thesis advisor, the eminent Larry Sandler often got quite aggravated when asked by a genetically naive colleague whether or not he thought it was "possible" to generate mutants of some given phenotype. "Of course, you can! You can get any mutant you want," Larry would bellow from his office, "I could mutate *E. coli* into an elephant, if given enough time." That may be stretching it a bit, but if you think enough about a given biological process, you can devise a screen for obtaining mutants in the process. It doesn't matter if such mutations are expected to be lethal, sterile, or whatever as homozygotes. We can cope with such things; it is done all the time. But a second question is more difficult. Can we make such mutants tell us what we want to know? Yes, but it is certain to require some serious thought. This will be the subject of the rest of this book.

2.1.2 Reason 2: To isolate more mutations in a specific gene of interest

As your analysis continues, you will likely focus your efforts on only a few of the genes defined by the mutations recovered in your general screen. In some cases, a specific gene might be important enough that you require multiple alleles of that

Box 2.3   A screen for sex-linked lethal mutations in Drosophila

Objective

To isolate recessive (usually loss-of-function) mutations in vital genes on the X chromosome. Embryos homozygous for such mutants might then be examined visually in search of any number of defects.

Basic stocks

Stock 1: *m/Y*. This stock consists of males carrying the recessive marker *miniature wings* (*m*) on the X chromosome mating to females homozygous for *m*.

Stock 2: *FM7*. The males in this stock carry both the X chromosome balancer FM7 and a normal Y chromosome. The balancer FM7c carries three overlapping inversions, and fully suppresses crossing over. It is marked with *yellow* (*y*), *white* (*w*), and *Bar* (*B*). *For those unfamiliar with the concept of balancer chromosomes, a review is provided in box 2.1. The females in this stock are homozygous for FM7.*

- The *y* mutant is fully recessive, and *y/Y* or *y/y* females display yellow bodies and yellow wings (normal flies are gray).
- The *w* mutant is fully recessive and *w/Y* males or *w/w* females display white eyes.
- The *B* mutant is dominant and both *B/Y* males and *B/+* females display abnormal kidney-shaped eyes.

Two features of the *FM7* chromosome make it an excellent balancer. First, the three inversions fully suppress crossing over. Second, the dominant *B* mutation allows you to follow this chromosome in stocks.

The screen itself

- Generation 0 (P1). Mutagenize *m/Y* males from Stock 1 by feeding them EMS, and then cross them to virgin *FM7/FM7* females from Stock 2.
- Generation 1 (F1). Pick up the *FM7/m** daughters. The symbol * denotes the mutagenized X chromosome. Mate these females to *FM7/Y*

males (either their brothers or more males from Stock 1). (You need not collect virgin females at this step, since females mating with their brothers is exactly the cross that we want to perform.) Place each female and three to five *FM7/Y* males (usually brothers) in a single vial.

- Generation 2 (F2). Examine the progeny of each vial. Look for vials that do not produce *m B*+ (miniature winged non-Bar) sons. Such vials must have been started with mothers heterozygous for a new recessive lethal mutant [denoted *l(1)n*]. That is to say that the sperm from that mutagenized male did indeed carry a newly induced lethal mutation.
- Generation 3 (F3). Take the *FM7/l(1)n m* daughters that did survive and cross them to their *FM7/Y* brothers. You have now established a stock of the new lethal mutant. You can use a variety of tricks to map the new mutant, to determine the time of death of the *m l(1)n* males, etc.

A complication: If you used a chemical mutagen and sample progeny derived from mature sperm, there is a good chance that many of your F1 females are in fact mosaics. If EMS got into a mature sperm and alkylated a given G, denoted G*, there are no subsequent replications to resolve the G*/C base pair. The alkylated base will remain opposite the C until the sperm fertilizes the egg and the first embryonic S phase commences. The result of that replication will be a G*–T daughter strand **and** a normal G–C daughter strand. When these two daughter chromatids segregate at the first embryonic mitosis, you have two genetically different populations of daughter cells.

If you imagine that the GC → AT transition indeed produced a recessive lethal, then you can see that this female has some cells bearing a new lethal mutation and some cells with a normal X chromosome. If that mosaicism extends to her germline, then the cells without the new lethal mutation will allow the female to produce miniature winged non-Bar progeny. You will discard this vial, and thus fail to recover this mutation.

Box 2.3 *(cont'd)*

There is a way to get around this problem, as shown below.

- Generation 0 (P1). Mutagenize *m*/*Y* males from Stock 1 by feeding them EMS, and cross them to virgin *FM7c*/*l(1)a* females.
- Generation 1 (F1). Pick up the *FM7c*/*m** females. The symbol * denotes the mutagenized X chromosome. Mate these females to *FM7c*/*Y* males. (Assure yourself that you need not collect virgin females at this step.) Place each female and three to five males in a single vial.
- Generation 2 (F2). Individually mate five to ten *FM7c*/*m** daughters from each female to *FM7c*/*Y* males. Place each female and three to five males in a single vial. This allows you to sample a number of progeny females from the original F1 female. If she is mosaic for a lethal, it is likely that some of the vials will produce no *m B*+ male progeny.

- Generation 3 (F3). Examine the progeny of each vial. Look for vials that do not produce *m B*+ (miniature winged non-Bar) sons. Such vials must have been started with mothers heterozygous for a new recessive lethal mutant [denoted *l(1)n*]. Mate the *FM7c*/*l(1)n m* sisters and cross to their *FM7c*/*Y* brothers.

In practice, many geneticists don't worry about mosaicism. First off, the best data suggest that only some 20–25% of the F1 will be mosaic in the germline. Second, the (cure described above is just too much work. (It involves increasing the number of single pair matings by five to ten times.) Given that most screens look at 10,000 to 20,000 single pair matings in the F1, a 100,000 vial F2 is a serious impediment to success. We are willing to accept the fact that we lose some fraction of newly induced mutations in this manner, so we just set up the original screen in a larger number.

---

gene, or you might require specific kinds of mutations in that gene, such as conditional alleles or null alleles. A second objective of screening for new mutations might therefore be to obtain new alleles at a specific gene. But the most common reason for wanting to isolate more mutations in a specific gene is the desire to have molecular "landmarks" for gene finding.

By gene finding I refer to the process of identifying a gene of interest in a stretch of already cloned DNA. As the various genome projects near completion, the most onerous hurdle faced by most geneticists is to determine just which base pairs of that DNA correspond to the gene of interest. Mutants associated with easily detectable alterations in DNA sequences, such as deletions or rearrangement breakpoints, prove invaluable in this analysis. The utility of such mutants reflects the problems inherent in trying to find a specific gene of interest within a region of known DNA sequence. Often, genetic mapping can position a gene of interest, such as a human disease gene, within a specific region of a DNA sequence whose boundaries are determined by nearby genetic markers that flank the gene of interest. But that region can sometimes contain tens or hundreds of transcription units, leaving the investigators wondering just which gene it is that when mutated creates the disease or phenotype of interest. A mutation that creates an easily detectable DNA rearrangement or sequence change in that gene provides a "signpost" that identifies the gene of interest.

2.1.3   Reason 3: To obtain mutation tools for structure–function analysis

There will come a point in your analysis when you know the sequence of the gene and its protein product. Perhaps the structure of that protein provides clues to its function. The next step is a high-resolution structure–function study of this protein to determine what functions are carried out by which amino acid domains. A large collection of missense mutations (mutations that result in the replacement of a single amino acid) now becomes an invaluable tool in your analysis, precisely because it allows you to examine the effects of changing the protein sequence "one amino acid at a time." One example of such a dissection, which involves the use of the *nanos* gene in Drosophila (Arrizabalaga & Lehmann 1999), will be discussed below.

2.1.4   Reason 4: To isolate mutations in a gene so far identified only by molecular approaches

Say, for example, that you have just discovered a Drosophila gene that has a sequence similar to that of an interesting human gene. In order to make this a useful result, you need mutants in this fly homolog, because until you demonstrate that those fly mutants have a phenotype similar to the human mutant, or a phenotype consistent with a defect in the predicted protein, all you have is sequence homology. You need to "knock out" that fly gene. The primary tool for specifically "knocking out" a given gene will be targeted gene disruption (see below).

   Regardless of your motive for making mutants, you want to be efficient about it. There are many means by which one can induce new mutations. Unfortunately, some people don't realize that the method they use will determine the sort of mutations they get. It is thus worth spending a page or so discussing these various tools for making mutants.

2.2   Mutagenesis and mutational mechanisms

Prior to the pioneering work of H. J. Muller in Drosophila, mutations were simply found by serendipitous accidents. They occurred spontaneously, and became the property of those individuals sharp-eyed enough to notice them and far-sighted enough to breed them. Unfortunately, the spontaneous mutation rate for most genes is low, approximately one new mutant per gene in every 100,000 to 1,000,000 gametes. Geneticists thus became dependent on either finding existing mutations within natural populations, or the rare recovery of new mutations in their laboratory strains. (For example, one of the greatest of all Drosophila geneticists, Calvin Bridges, made his first contribution to genetics by noticing a fly with an unusual eye color, vermillion, among the flies buzzing in a bottle that he was about to wash. That fly turned out to carry an incredibly useful new eye color mutation.)

One of the fundamental advances of twentieth century genetics was the finding that a variety of physical, chemical, or biological agents could induce mutations. Muller won the Nobel Prize for demonstrating that X-rays induced mutations. Studies carried out later in the century would reveal the variety of chemicals that are also potent mutagens. Transposable elements, long known as mutator elements by corn geneticists, were recognized as powerful mutators in a variety of organisms by the 1980s, and the technology of *in vitro* DNA manipulation quickly allowed transposable elements to be modified in such a way as to facilitate cloning of the mutant genes.

Each of these agents provides a powerful means for inducing new mutations. However, they differ in the types of mutations they induce, the frequency with which they induce mutations, and the genes they can efficiently mutate. There follows a comparison of the advantages and disadvantages of the three major classes of mutagens.

2.2.1 Method 1: Ionizing radiation (usually X-rays and gamma-rays)

The passage of ionizing radiation through a cell can damage DNA by both direct and indirect means. Radiation interacts directly with DNA, but DNA can also be damaged as a secondary consequence of the radiation interacting with the water in the cell. The reaction of water with free radicals, for example, can result in the production of DNA-damaging chemicals. The resulting damage includes damage to the nitrogenous bases, as well as breakage of the sugar–phosphate backbone (strand breakage).

Because the transfer of energy as a beam of radiation passing through the cell is not uniform, but rather reflects a series of discrete energy depositions, damage to the DNA due to radiation is often clustered within small regions on the DNA. If two single-strand breaks occur sufficiently close to each other, but on opposite strands of the DNA duplex, a double-strand break will occur. Single- and double-strand breaks most likely account for much of the lethal effect of ionizing radiation. Most critically, the ends of the DNA at radiation-induced breaks are often chemically damaged in such a way as to preclude simple re-ligation. The types of mutations obtained following ionizing radiation often reflect loss or addition of base pairs or gross chromosome rearrangements. Accordingly then, X-rays or gamma-rays are not good tools for making missense mutations, but are excellent tools for making deletions and/or frameshift mutations.

Indeed, ionizing radiation is probably the best tool available for producing chromosome aberrations, and it is one of the best methods for making both small and large deletions. Unfortunately, ionizing radiation is not an efficient mutagen for most genes. The "rule of thumb" that most Drosophila geneticists use is one to two new mutants in a given gene per 10,000 gametes following a dose of approximately 3,500 R. As an example, Zhang and Hawley (1990) recovered nine new loss-of-function alleles of the Drosophila *nod* gene from among 110,000 treated chromosomes following treatment of *nod*+-bearing males with 3,500 R. This included one large deletion and one translocation event. There were six mutations that created small deletions within the gene and one, presumably spontaneous, transposon insertion. [For a more detailed discussion of the relationship between dose rate and mutation rate when using radiation, see Ashburner (1989).]

If one needs breakpoints or deletions to facilitate some molecular objective (such as gene finding within a cloned region of DNA), ionizing radiation is often the mutagen of choice. You are probably better off with chemical mutagens if you need a large number of new mutants in a given gene or set of genes.

2.2.2   *Method 2: Chemical mutagens*

Chemical mutagenesis was developed by Charlotte Auerbach and her collaborators in England prior to and during the Second World War as part of a chemical warfare program. The value of chemical mutagenesis is that it can be far more efficient than ionizing radiation. *The drawback is that it can be risky for the investigator!*

We will divide our consideration of chemical mutagens into two groups, namely alkylating agents and cross-linking agents. The commonly used alkylating agents, such as ethylmethane sulfonate (EMS) or methlymethane sulfonate (MMS), introduce small alkyl groups ($-CH_2CH_3$ or $-CH_3$) onto the bases themselves (figure 2.1), thus modifying their capacity for Watson–Crick base pairing. An alkylated guanine pairs with thymine, rather than with cytosine. Thus, if this lesion is not repaired before the next replication, a T will be incorporated in place of a C. Following the second replication event, the original GC base pair will be replaced by an AT, resulting in a GC → AT transition. (By tradition GC → AT and AT → GC mutations are referred to as *transitions*, while AT → TA, AT → CG, TA → GC, and TA → AT changes are referred to as *transversions*.) Modification of thymine by some alkylating agents can also produce TA → CG

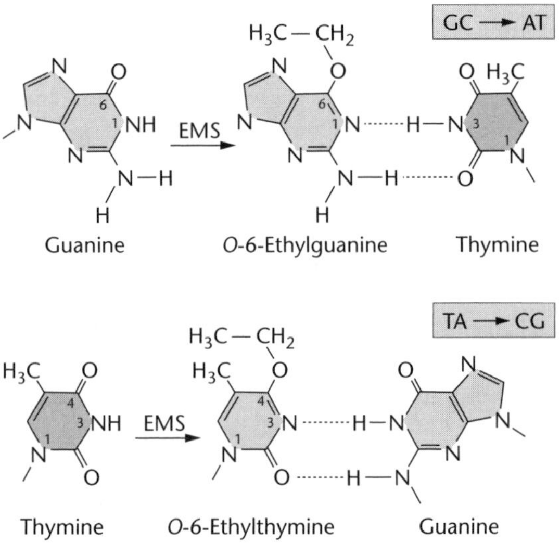

Figure 2.1   The commonly used alkylating agents, such as EMS, introduce small alkyl groups ($-CH_2CH_3$) onto the bases themselves

Alkylehng

transitions by a similar mechanism. Accordingly, a substantial fraction of the mutants produced by EMS mutagenesis are either missense or nonsense mutations. EMS does produce more complex types of mutations, such as deletions, albeit at a lower frequency. The mechanism by which such aberrations arise is not understood.

As an example, Arrizabalaga and Lehmann (1999) used a clever selective screen to isolate EMS-induced mutants in the *nanos* gene in Drosophila. The screen targeted a 1,661 base pair *nanos* transgene construct whose overexpression of *nanos* caused female sterility. Loss-of-function mutants in the *nanos* gene carried by this construct restored fertility. The strategy was to mutagenize transgene-bearing males, cross them to appropriate females, and recover the daughters bearing the mutagenized transgene. (The actual details of this series of crosses are described below.) From among 186,000 daughters, they recovered 68 fertile females, all of which were shown to be the consequence of mutations of the *nanos* transgene (for a rather low mutation frequency of 3.6×10^{-4} per gene per generation). Of these new alleles, 60 were shown to be the consequence of mutations within the protein-coding sequence. All but two of these mutations affect a single base. Of the 58 single base mutations, 50 were due to the expected $GC \rightarrow AT$ transitions. Six others were the result of $AT \rightarrow TA$ transversions and two were the result of $TA \rightarrow GC$ transversions. Among these 58 single base pair changes, 27 created new stop codons. The remaining mutations were either missense alleles (28) or mutations that affect the intron/exon structure of the gene (3). The two mutations that affected more than one base were a three base pair in-frame deletion and a 13 base pair insertion.

The real advantage of using alkylating agents is the very high frequency with which they induce mutations. Again, the chemical mutagenesis rule of thumb in Drosophila is that the EMS-induced per gene mutation rate is 1 in 500 to 1 in 1,000 treated gametes. Thus, the yield using alkylating agents can be 10–20 times greater than that obtained with ionizing radiation.

← high frequency

The second class of chemical mutagen is the cross-linking agents. Cross-linking agents such as diepoxybutane (DEB) and *cis*-platinum (II) diamino-dichloride (*cis*-platinum) can chemically interlock two sites on the DNA molecules. Cross-links that connect two sites on the same strand are known as intra-strand cross-links, and those that connect opposite strands are referred to as inter-strand cross-links. Both types of cross-links prevent replication and are difficult for the cell to repair. When incorrectly repaired, such lesions often produce small deletions. Cross-linking agents can be efficient mutagens, but their use is sometimes limited by their high toxicity.

cross link agent

small deletion

There are two major drawbacks to the use of chemical mutagens. First, most chemically induced changes are detectable only by direct sequencing of the gene. For this reason, most, but not all, chemically induced mutations are of little use in the process of gene finding. Second, as noted above, these chemicals are dangerous. Before ordering a mutagen make sure you understand how to handle it. (For example, EMS is shipped in a rectangular cardboard box. Inside the box, the chemical is stored in a glass bottle wrapped in a corrugated cardboard tube. The tube has no bottom. We have seen a new invest-

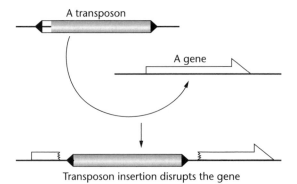

Figure 2.2   The insertion of the transposon creates a mutant

igator pull the bottle out of the box by grabbing the cardboard tube. Upon exiting the box, the corrugated tube expanded and the bottle fell to the floor. Fortunately, very fortunately, it was caught before it shattered.

2.2.3   Method 3: Transposons as mutagens

If one is searching for mutations in order to assist in gene cloning, the best possible tool is transposons. A decade or so ago Barbara McClintock rather belatedly won the Nobel Prize for her work on transposon movement in corn. In the intervening years, transposon-based mutagenesis has become a standard tool for making mutants in flies, worms, fungi, etc. The insertion of the transposon creates a mutant (see below) and, provided one has a way to identify the transposon DNA, also creates a means to recover the sequences flanking the insertion site of the transposon (figure 2.2).

The basic premise of transposon-based mutagenesis is that the insertion of a transposon into a given gene will disrupt the function of that gene. The mechanism by which the disruption can occur is varied, and that variation itself can create useful tools. The basic mechanisms for creating loss-of-function mutations are:

1   Disruption of the coding sequence itself.
2   Disruption of a crucial regulatory sequence.
3   Blocking of enhancer–promoter interactions.
4   Creation of improper splice or termination sequences.

However, properly engineered transposons can also be used to create gain-of-function mutants. When these insertions carry with them strong enhancer elements they can serve to activate neighboring genes in a tissue-specific fashion. For example, one could create a library of insertions in Drosophila bearing the yeast enhancer element, known as a UAS element. This enhancer can be activated by inserting a copy of the GAL4 transcriptional activator protein, which binds the UAS sequence, into the fly under the control of a regulatable promoter. The resulting GAL4–UAS interaction will serve to activate the expression of those genes near the UAS insertion at those times or in those tissues where GAL4 is being expressed. Alternatively, one could create a library of insertions bearing enhancerless GAL4 genes, whose expression will be controlled by enhancers

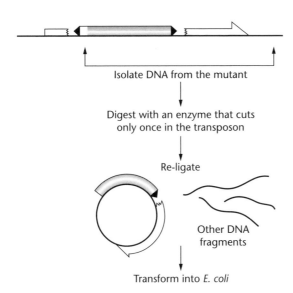

Isolate DNA from the mutant

Digest with an enzyme that cuts
only once in the transposon

Re-ligate

Other DNA
fragments

Transform into *E. coli*

Figure 2.3 Provided that the transposon carries sequences that allow replication and selection in bacteria, one can thus re-clone the transposon and flanking DNA sequences by a technique known as plasmid rescue

mapping near the insertion site. This library can be tested by crossing in a useful reporter gene (e.g. lacZ or a GFP construct) under the control of a UAS element. This "enhancer trapping" technique allows you to find enhancers (and thus genes) that function in a specific time and or place during development. If you are lucky the genes you find will have also been disrupted by the same Gal4-bearing insertions that were used to identify them in the first place.

Controlling transposon movement: transposons as a tool have become more valuable as geneticists develop better tools to control the process of transposition. In Drosophila, the transposon of choice is usually a P element. Mobilization occurs only when a P element-encoded enzyme called transposase is produced. This enzyme acts at the ends of the P element to carry out the transposition process. The presence or absence of transposase in a germline can be controlled by using a crippled P element. Although this crippled element cannot transpose itself, the enzyme it produces can cause the transposition of other "target" P elements, including elements that lack their own transposase gene but do have an intact end sequence. By controlling the presence or transcriptional activity of this crippled P element, one can determine exactly when movement of target transposons will be allowed. These target transposons can move in the presence of transposase, but cannot produce it themselves. Thus, they move only when they and the crippled master element share a genome.

Target P elements can carry marker genes that confer a visible phenotype upon the fly, allowing these elements to be followed in crosses. These elements also carry a bacterial origin of replication, unique restriction sites within the transposon, and a selectable antibiotic-resistant gene. One can thus re-clone the transposon and flanking DNA sequences by a technique known as plasmid rescue (figure 2.3). Briefly, genomic DNA is isolated from individuals carrying the new insertion, and digested with a restriction enzyme that cuts only once in the transposon. The DNA is then ligated at a low enough concentration so that

You can control transposase

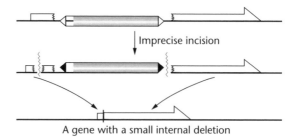

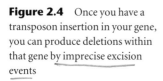

Figure 2.4 Once you have a transposon insertion in your gene, you can produce deletions within that gene by imprecise excision events

the ends of each DNA molecule are ligated to form circles. The ligated mix is transformed into *E. coli*. Following antibiotic selection, the resulting colonies carry plasmids bearing the transposon and the DNA sequences of your gene that flank the site of P element insertion. Thus, if you can mobilize a P element into your gene, you have at least part of your gene cloned.

One real value of transposon mutagenesis is that it exploits the penchant of many transposons to move to new sites that are close to the original insertion. Thus, if one already has a transposon inserted near to, but not within, the gene of interest, one can often easily mobilize that transposon into the gene of interest. By doing so, one can subsequently obtain flanking DNA sequences corresponding to that gene. A second value of transposon mutagenesis is that excision is often, or can be made to be, an imprecise process that leads to deletion of flanking DNA sequences. Thus, once you have a transposon insertion in your gene, you can produce deletions within that gene by imprecise excision events (figure 2.4).

The reader might then ask the question: "Why, in the name of Mendel, would one mess with the X-rays or the deadly chemicals? Who wants to finger-paint with carcinogens when you get the mutant **and** the gene cloned using transposon mutagenesis?" That reader might have a point. But there are reasons to be just a bit shy of these screens. The primary difficulty with doing a transposon mutagenesis is that most transposons show some degree of target site specificity. Thus, they insert efficiently into some genes while ignoring others. There are genes that are very mutable with some transposons, and others that are hit rarely, if at all. Additionally, some transposons prefer regulatory regions or introns as their target sites; insertion into these regions produces only a partial decrease in gene activity. One can get around these difficulties, but it requires additional effort.

So which mutagen should you use? Part of the answer depends on where you start. In general, our advice is to start with EMS. An EMS mutagenesis will tell you what genes are out there, and give you some idea of what kind of phenotypes you can expect. Then you can use ionizing radiation or transposon mutagenesis to mutate those genes that most interest you. All three methods work, and unless you are luckier than most of us, you will probably need to use all three.

2.2.4   Method 4: Targeted gene disruption (a variant on transposon mutagenesis)

Some organisms are now amenable to targeted gene disruption. In this case one can directly insert (via transformation) an exogenous DNA fragment into the gene of interest. One can either insert foreign DNA into the gene or replace sequences within the gene with some desired sequences. Targeted disruption relies on the homologous integration of transformed DNA molecules to disrupt or alter the gene of interest. The critical issue here is that you need to already possess a clone of the gene of interest, and you must be working in one of the few organisms where targeting works (e.g. *E. coli*, yeast, Drosophila, mice, and human cell lines). The value of disruption is that one can construct null alleles and recover them without having to know their phenotype in advance.

We should also note that in cases where one has the cloned gene in hand, and wants to make mutants, it is often possible to mimic the effects of such mutants by using techniques known as RNAi or co-suppression. These techniques are summarized in box 2.4.

Box 2.4   Making phenocopies by RNAi and co-suppression

If one has a cloned gene and wishes to know the consequences of ablating the function of that gene, there are two possible approaches to produce phenotypic mimics, or **phenocopies**, of a loss-of-function mutant. The first technique, known as RNA interference (RNAi) involves the injection of double-stranded RNA (dsRNA) homologous to the gene of interest into the animal, tissue, or cell. The second technique, known as co-suppression, involves the introduction of homologous double-stranded DNA into the organism or tissue. In both cases the double-stranded nucleic acid acts to disrupt expression of the homologous gene post-transcriptionally, via an RNA intermediate. For example, introduction of dsRNA into an adult worm results in the destruction of the homologous mRNA molecules in both the adult and its progeny. In doing so, the dsRNA injection "phenocopies" the effect of a strong hypomorphic or nullomorphic mutant. In organisms and systems where it works, it can be used as a tool to determine what the phenotype of a mutant in a given gene is likely to be, and thus allow one to devise screens or selections for that type of mutant. One can also use this technique to determine whether or not a given gene plays a critical role in some process of interest.

Co-suppression is well documented in plants, fungi, ciliates, and the *C. elegans* germline [for a review, see Dernburg et al. (2000)]. RNAi is well documented in *C. elegans* [for a review, see Hsieh and Fire (2000)], Drosophila [tissue culture cells (Caplen et al. 2000); whole animals (Kennerdell & Carthew 1998, 2000)], mammalian cell lines (Hope 2001), including mouse ES cells and mouse embryos (Svoboda et al. 2001; Yang et al. 2001), and plants (Waterhouse et al. 1998; Chuang & Meyerowitz 2000; Escobar et al. 2001). Indeed, RNAi works so well in *C. elegans* that one doesn't even need to inject the worms, one can just soak them in a solution containing the bacteria expressing the dsRNA, and let them eat the bacteria (Timmons et al. 2001).

The exact mechanisms by which RNAi and co-suppression work remain unclear [cf. Hunter (2000) and Carthew (2001)], although there is substantial evidence that the two processes share some, but not all, components (Dernburg et al. 2000; Maine 2000). Efforts to work out the mechanisms are in progress, and seem likely to yield fruit quite soon (Grishok et al. 2001; Parrish & Fire 2001). But regardless of how these processes work, the fact is that they do work, and thus they provide a quick and usually easy method to make phenocopies.

2.3   What phenotype should you screen (or select) for?

The success of this effort will depend, as we will reiterate constantly, on how cleverly you set up your initial screening criteria. The phenotype of the mutants you select will determine both the genes you identify and the types of mutations you recover. No aspect of a mutant hunt is as critical as determining the phenotype on which the screen will be based.

Suppose, to follow up the flight example presented above, we search for mutants in Drosophila on the basis of their inability to fly. As long as our screen is based on testing *live* animals for an inability to fly, we will obtain only those mutants that are not themselves lethal. Thus, we are likely to miss those genes whose protein products are essential to the organism's normal development or survival, in addition to being required for flight. This may be fine, perhaps even desirable, in some circumstances, but it may severely limit the value of the mutant collection that you obtain. Similarly, there may be other genes that encode subtle components of flight (e.g. navigation), whose effects might be missed by a simple "does it fly, or not?" screen.

Focus on too narrow or demanding a phenotype, and you may miss a great many genes. Cast your net too widely (e.g. trying to identify *any* mutation that is homozygous lethal or sterile) and you will be swamped by mutants of genes that may be of little interest to you. On this matter we can offer only three pieces of advice.

First, start by looking for a phenotype that is specific to the process in which you are interested. In other words, if you are interested in flight, look for flight-less mutants. Don't expand your screen to "any gene that can mutate to a lethal" just because flight may require some vital genes. Don't worry initially about missing some genes; no screen is perfect. Once you obtain several good mutants, they can usually serve as tools to help you find mutants in the sorts of genes that your initial screen may have missed. This is not to say that there isn't an enormous value for huge screens designed to recover lethal mutations in every vital gene, or sterile mutants in every gene required for fertility. Such screens, when done cooperatively and made available to the entire community, are of enormous value. But they are usually not a good way for an individual scientist to invest her or his time, especially if they are interested in a specific biological process.

Second, to the best of your ability, choose a clean, discrete, and easy to score phenotype. Helen Francis-Lang of Exelexis Inc., a biotech company, refers to a mutant screen-friendly phenotype as a "screenotype." It is a useful term, reminding us that there may be phenotypes that are "scorable" in any given experiment but may not be useful as an endpoint in large-scale screening. Because mutant hunts are messy and time-consuming affairs, you need to be able to take a quick look at each individual and reliably determine its phenotype quickly and unambiguously. Selecting for a simple visible phenotype is usually best. As difficult as it might be to imagine, it is often possible to devise screens for a given biological process that are based on an externally visible phenotypic change. For example, Gerry Rubin dissected the process of signal transduction in the developing eye of the fruit fly by using screens in which the endpoint was the texture of the adult eye (Simon et al. 1991; Rebay et al. 2000). A mutant in a signal transduction protein was found that changed the eye texture of the fly. By screening for mutants that enhanced or suppressed this phenotype, Rubin obtained mutants in other genes whose protein products participate in this pathway.

In some cases, the observable phenotype, while process-specific, need not be obviously related to the process under study. Gary Karpen obtained a fascinating collection of mutants that affect the function of centromeres in Drosophila isolated entirely by screens in which the sought after change in phenotype was a change in the eye and body color of the fly (Murphy & Karpen 1995). Karpen created small "marker" chromosomes in which the genes responsible for eye and body color were moved close to the centromeric regions. He then screened for deletions of these markers, some of which extended into the centromeric regions. Just as in the case discussed above, Karpen developed a genetic system in which changes in chromosome structure, or the signal transduction apparatus, resulted in changes in an easily observed visible phenotype. Simply put, the biological system you are studying can be as complex as you wish, but your screens need to be as simple as possible.

Third, devise a scheme in which organisms displaying the mutant phenotype are selected for you in some fashion, rather than using a screen where one just searches through all the progeny looking for mutants. If the mutants you want are the only progeny that get to live, then such selections are much less work than most screens. In my laboratory such mutant selection schemes are referred to as "mutate my way, or die!" experiments. The drawback to such screens is that the tighter you make your selection scheme, the fewer types of mutations you recover.

2.4 Actually getting started

2.4.1 *Your starting material*

A colleague of ours tells a sad story about doing a huge screen for behavioral mutants in flies only to recover the same odd mutation 20 or more times. It turns out that this unusual allele pre-existed, albeit at a low frequency, in the starting stocks. We cannot urge you strongly enough to check multiple isolates of your starting stock for the phenotype you are about to assay. Better yet, take the time to build an isogenic starting stock from single pairs of individuals you know to be wildtype for the phenotype. The old maxim "garbage in–garbage out" truly applies here. (An isogenic stock is comprised of an individual homozygous for all the loci on a designated chromosome or set of chromosomes.)

2.4.2 *Pilot screens*

There are many screens that work just great on blackboards and fail miserably on fly or yeast media. Trust us, however you design your first screen, it probably won't work exactly the way you thought. Some genotype won't survive, some class of females won't be fertile enough, etc. And the more cute tricks you designed into your screen, the more things can go wrong. So, do a small pilot screen to make sure your ideas work in real life.

2.4.3   *Keeping too many, keeping too few*

There is a tendency in the first phase of a large screen to keep everything that looks odd, no matter how subtle the difference. Remember to choose a strong, clean "screenotype" and stick with it. This is especially true if your phenotype is to any degree quantitative. Choose your "this is interesting" cut-off point rigidly before you start, or after your pilot screen, and stick to it.

2.4.4   *How many mutants is enough?*

Okay, so you've been through 20,000 or 200,000 mutagen-treated chromosomes to screen for new mutants in process X. You have 40 or 400 new mutants. Should you stop? The critical issue here, unfortunately, is not how many mutants you have, but how many mutants you don't have. You are interested in estimating the fraction of the total number of genes that can mutate to the phenotype you seek, which are represented by your current collection. If, for example, you have 100 mutants, with 20 alleles defining each of five genes, then for heaven's sake stop! You can keep this up for years and not identify any more new genes. A plot of the numbers of new genes identified against the number of mutants recovered was generated by Nusslein-Volhard et al. (1984) and is displayed in figure 2.5. One can see that the identification of more and more genes just as a consequence of finding more and more new mutants is by no means a sure thing. The Law of Diminishing Returns certainly applies here. On the other hand, if you have 40 mutants, with 20 alleles of one gene and only one or two alleles of some 10–15 other genes, then no, you have probably not yet reached saturation. There are still likely to be genes that aren't represented by mutants in your collection. You *might* want to keep going.

   So how can you estimate the number of genes not represented by mutants in your new collection? To cope with this problem, geneticists have agreed by convention to lie to themselves. If we assume that all loci are *equally* mutable at some low frequency, we can use the Poisson distribution to estimate the fraction of pre-

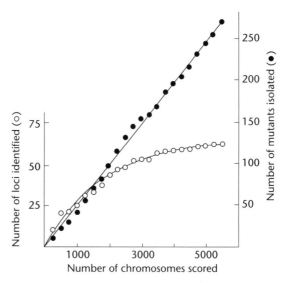

Figure 2.5   A plot of the numbers of new genes identified, and the number of mutants recovered, against the number of mutagen-treated lines tested in the screen performed by Nusslein-Volhard and collaborators

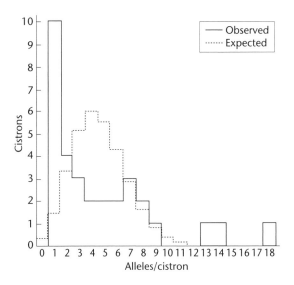

Figure 2.6 The observed and expected distributions of *cdc* alleles per cistron for the Hartwell et al. (1973) screen (based on an average of 4.625 alleles per cistron from the Poisson distribution). (Adapted from Hartwell et al. 1973)

sumably mutable genes for which no alleles have yet been recovered ($f[0]$). For a screen in which the mean number of alleles per loci already defined is m, then:

$$f[0] = e^{-m}$$

Our rule of thumb is that saturation is achieved when $m = 5$ and thus $f[0] = 0.007$. That's fine. It is a total delusion for two reasons, but it is fine. Firstly, this calculation is a delusion because we all know that some loci are far more mutable than others, and thus one should not use the Poisson distribution. Indeed, the fact that some loci are far more mutable than others goes back to Morgan and Muller [cited in Lefevre and Watkins (1986)]. Secondly, and of equal concern, is the fact that the mean value (the essence of using the Poisson distribution) is actually being miscalculated in this equation. A true estimation of the mean would require knowing the number of loci represented by zero alleles, which is of course exactly what one is trying to calculate. This error will greatly inflate the estimation of the mean (m), and thus underestimate the number of loci predicted to be represented by smaller numbers of alleles, especially the number of loci for which no alleles were obtained.

Figure 2.6 presents the observed distribution of alleles per locus for the Hartwell et al. (1973) screen. It is clear that in this case, the fit to the Poisson distribution is relatively poor. There are both too many loci defined by only one allele and too many defined by large numbers of alleles. These are not unusual results. My laboratory, in collaboration with Dr. Ken Burtis, just did a large screen for DNA repair-defective mutants in flies. After screening approximately 6,000 EMS-treated third chromosomes, we recovered 69 new mutants. But 25 of these mutants were shown to be alleles of the same (very mutable) gene. The remaining 44 mutants defined genes that were represented by either one (22 cases), two (three cases), four (one case), five (one case), or seven (one case) new alleles. If we just take the data at face value then the average number of alleles per gene is approximately 69/29, or 2.38, and thus $f[0] = 0.093$. If we believe those data, then we have identified more than 90% of the genes that can be identified in such a screen. That really is more than enough.

But seriously, these data don't fit a Poisson distribution. Clearly, that gene with 25 alleles is a lot more mutable than the other genes. Either the mutation that was recovered 25 times pre-existed in the stock prior to mutagenesis (a rude possibility that most people wouldn't consider, but a possibility none the less) or that gene is hypermutable under these conditions. Either way, these alleles cannot be used to assess the degree of saturation. So what if we repeat the calculation leaving out the over-represented gene. In this case, the average number of alleles per gene is approximately 44/28, or 1.57, and thus $f[0] = 0.208$. This calculation still suggests that we have close to 80% of the genes one can get by this method. None the less, the mathematics argues that the screen is by no means saturated. Moreover, had we screened another 2,000 chromosomes (and not identified a new complementation group), we would still only be confident that we had 90% of the genes. Hardly a worthwhile effort. So what does one do? The problem is that to continue turning the crank and screening more mutants may not be worth the effort (see figure 2.5). In our case we decided that enough was enough, and stopped screening.

In general, despite the flaws heralded above, this misapplication of the Poisson distribution can be considered a reasonable approximation as long as the mean number of alleles per locus is large. We should note that other statistical approaches for estimating the degree of saturation (including using truncated binomial expansions and gamma distributions) have been proposed and tested [for a review, see Lefevre and Watkins (1986)]. The distributions predicted by these methods produce better, though far from perfect, fits to the observed distributions.

But pragmatically, if you already have an average three mutants for each gene, then the effort to find a much less represented gene is probably just not worth the work. The answer then is: you can stop at $m = 5$ and no one will fault you. You can also stop at $m = 3$, you just cannot say that your screen was saturating. In reality, you will stop at the point where you simply cannot, or the people in your laboratory will not, keep on going. At some point they will decide to invest their time in characterizing the mutants in hand, rather than making more mutants.

Summary

This chapter was intended to describe three issues: why you might wish to screen for mutants; how you might choose to induce mutations; and how to actually execute the screen. We chose several examples of both screens and selections, but our focus was always the same: to make it clear that the kinds of mutants you recover are dictated by the type of screen you use. We noted that each type of screen or selection has limitations, no screen will get "every mutant." Thus, we noted a need to carefully choose the initial strain to mutagenize, the mutagen to use, and the "screenotype" to be hunted for or selected. The more carefully designed, and tested, a screen is, the more likely it is to work. Still, we hope we didn't leave you with the impression that even the most careful of designs can predict all of the outcomes. Sometimes the most interesting mutants are the ones you never expected to get.

Box 2.5 Reviews of mutant isolation schemes and techniques in various organisms

Prokaryotes and their viruses

Phage lambda

1 Davis, R. W., Botstein, D., and Roth, J. (1980). *Advanced Bacterial Genetics Laboratory Manual.* Cold Spring Harbor Laboratory Press, Cold Spring Harbor, NY.
2 Hendrix et al. (eds.) (1983). *Lambda II.* Cold Spring Harbor Laboratory Press, Cold Spring Harbor, NY. See especially the chapters by Arber (A beginner's guide to lambda biology) and Arber et al. (Experimental methods for use with lambda).

Phage T4

1 American Society of Microbiology website: http://www.asmusa.org/division/m/M.html.
2 Calendar, R. (1988). *The Bacteriophages.* Plenum Press, New York, two volumes.
3 Mathews, C. K. and the American Society for Microbiology (1983). *Bacteriophage T4.* ASM, Washington, DC.
4 Phage Ecology Group (terrific images!): http://www.phage.org/beg_phage_images.html.
5 The Bacteriophge Home Page: http://www.evergreen.edu/phage.

Phage P22

1 Casjens, S. American Society of Microbiology P22 website: http://www.asmusa.org/division/m/fax/P22Fax.html.
2 Susskind, M. and Botstein, D. (1978). Molecular genetics of bacteriophage P22. *Microbiol Rev* 42: 385–413.

E. coli

1 Davis, R. W., Botstein, D., and Roth, J. (1980). *Advanced Bacterial Genetics Laboratory Manual.* Cold Spring Harbor Laboratory Press, Cold Spring Harbor, NY.

2 Miller, J. H. (1992). *A Short Course in Bacterial Genetics: A Laboratory Manual and Handbook for* Escherichia coli *and Related Bacteria.* Cold Spring Harbor Laboratory Press, Cold Spring Harbor, NY.
3 The *E. coli* Genetic Stock Center: http://cgsc.biology.yale.edu/.

Salmonella typhimurium

1 Davis, R. W., Botstein, D., and Roth, J. (1980). *Advanced Bacterial Genetics Laboratory Manual.* Cold Spring Harbor Laboratory Press, Cold Spring Harbor, NY.
2 Sanderson, K. E. and Roth, J. R. (1983). The linkage map of *Salmonella typhimurium. Microbiol Rev* 47: 410–53.
3 Sanderson, K. E. and Roth, J. R. (1988). The linkage map of *Salmonella typhimurium.* Edition VII. *Microbiol Rev* 52: 485–532.
4 The Salmonella Stock Center: http://www.salmonella.org/.

Caulobacter

1 Ely, B. (1991). Genetics of *Caulobacter crescentus. Methods Enzymol* 204: 372–84.

The following is a list of more modern articles dealing with current approaches to genetic analysis in a variety of bacteria.
1 Akerley, B. J., Rubin, E. J., Camilli, A., Lampe, D. J., Robertson, H. M., and Mekalanos, J. J. (1998). Systematic identification of essential genes by in vitro mariner mutagenesis. *Proc Natl Acad Sci USA* 95: 8927–32.
2 Biery, M. C., Stewart, F. J., Stellwagen, A. E., Raleigh, E. A., and Craig, N. L. (2000). A simple in vitro Tn7-based transposition system with low target site selectivity for genome and gene analysis. *Nucleic Acids Res* 28: 1067–77.
3 Braunstein, M., Griffin, T. J. IV, Kriakov, J. I., Friedman, S. T., Grindley, N. D., and Jacobs,

Box 2.5   (*cont'd*)

W. R. Jr. (2000). Identification of genes encoding exported *Mycobacterium tuberculosis* proteins using a Tn552′phoA in vitro transposition system. *J Bacteriol* 182: 2732–40.

4   Conner, C. P., Heithoff, D. M., and Mahan, M. J. (1998). In vivo gene expression: contributions to infection, virulence, and pathogenesis. *Curr Topics Microbiol Immunol* 225: 1–11.

5   Shea, J. E. and Holden, D. W. (2000). Signature-tagged mutagenesis helps identify virulence genes. *ASM News* 66: 15–20.

6   Zhang, L., Foxman, B., Manning, S. D., Tallman, P., and Marrs, C. F. (2000). Molecular epidemiologic approaches to virulence gene discovery in uropathogenic *Escherichia coli*. *Infect Immunol* 68: 2009–15.

Single-celled eukaryotes: yeasts, other fungi, and algae

S. cerevisiae

1   DeRisi, J. L., Iyer, V. R., and Brown, P. O. (1997). Exploring the metabolic and genetic control of gene expression on a genomic scale. *Science* 278: 680–6.

2   Forsburg, S. L. (2001). The art and design of genetic screens: yeast. *Nat Rev Genet* 2(9): 659–68.

3   Guthrie, C. and Fink, G. R. (eds.) (1991). *Methods in Enzymology, Vol. 194, Guide to Yeast Genetics and Molecular Biology*. Academic Press, New York.

4   http://genome-www.stanford.edu/Saccharomyces/ (the core database).

5   Jones, E. W., Pringle, J. R., and Broach, J. R. (eds.) (1992). *The Molecular and Cellular Biology of the Yeast* Saccharomyces, *Vol. 2, Gene Expression*. Cold Spring Harbor Laboratory Press, Cold Spring Harbor, NY.

6   Sherman, F. A superb tutorial in yeast genetics and molecular biology: http://dbb.urmc.rochester.edu/labs/Sherman_f/yeast/1.html.

7   Sherman, F. (1991). Getting started with yeast. In: C. Guthrie and G. R. Fink (eds.), *Methods in Enzymology, Vol. 194, Guide to Yeast Genetics and Molecular Biology*. Academic Press, New York, pp. 3–21.

Schizosaccharomyces pombe

1   Forsburg, S. L. (2001). The art and design of genetic screens: yeast. *Nat Rev Genet* 2(9): 659–68.

2   Forsburg, S. (2002). Pombe website: http://pingu.salk.edu/~forsburg/pombeweb.html.

3   Hayles, J. and Nurse, P. (1992). Genetics of fission yeast *Schizosaccharomyces pombe*. *Ann Rev Genet* 26: 373–402.

4   Hochstenback, F. (1999). *The Fission Yeast Handbook*: http://www.bio.uva.nl/pombe/handbook/.

5   Moreno, S., Klar, A., and Nurse, P. (1991). Molecular genetic analysis of fission yeast *Schizosaccharomyces pombe*. *Methods Enzymol* 194: 795–823.

6   The Forsburg Laboratory website: http://pingu.salk.edu/~forsburg/lab.html.

Aspergillus nidulans

1   Kaminskyj, S. G. W. Fundamentals of growth, storage, genetics and microscopy of *Aspergillus nidulans*: http://www.hgmp.mrc.ac.uk/research/fgsc/fgn48/Kaminskyj.html.

2   The Aspergillus website: http://www.gla.ac.uk/Acad/IBLS/molgen/aspergillus/index.html.

3   The Fungal Genetics Stock Center: http://www.fgsc.net/.

Chlamydomonas reinhardtii

1   Dutcher, S. K. (1995). Mating and tetrad analysis in *Chlamydomonas reinhardtii*. *Methods Cell Biol* 47: 531–40.

2   Harris, E. (1989). *The Chlamydomonas Source Book: A Comprehensive Guide to Biology and Laboratory Use*. Academic Press, New York.

3   The Chlamydomonas Stock Center and Genetics Resource:http://www.biology.duke.edu/chlamy/.

Dictyostelium

1   Kessin, R. H. (2001). *Dictyostelium: Evolution, Cell, Biology, and the Development of Multicellularity*. Cambridge University Press, Cambridge, UK.

2   Kuspa, A. and Loomis, W. F. (1994). Transformation of *Dictyostelium*: Gene disruptions,

insertional mutagenesis, and promoter traps. *Methods Molec Genet* 3: 3–21.

3   Scherczinger, C. A. and Kenetch, D. A. (1993). Co-suppression of *Dictyostelium discoideum* myosin II heavy-chain gene expression by a sense orientation transcript. *Antisense Res Dev* 3(2): 207–17.

4   The Dicty website: http://dictybase.org/.

Neurospora crassa

1   Davis, R. H. and de Serres, F. J. (1970). Genetic and microbiological research techniques for *Neurospora crassa*. *Methods Enzymol* 27A: 79–143.

2   Davis, R. H. (2000). *Neurospora: Contributions of a Model Organism*. Oxford University Press, New York (see especially Chapter 14: Genetic, biochemical, and molecular techniques).

3   Ellis, C. H. (2001). *Neurospora and Tetrad Analysis*: http://www.dac.neu.edu/biology/c.ellis/gen11/.

4   Perkins, D. D. and Barry, E. G. (1977). The cytogenetics of *Neurospora*. *Adv Genet* 19: 133–285.

5   Perkins, D. D. (1997). Chromosome rearrangements in *Neurospora* and other filamentous fungi. *Adv Genet* 36: 239–398.

6   Perkins, D. D., Radford, A., and Sachs, M. S. (2000). *The* Neurospora *Compendium: Chromosomal Loci*. Academic Press, San Diego, CA.

7   The Fungal Genetics Stock Center: http://www.fgsc.net/.

Sordaria

1   Fields, W. G. (1970). An introduction to the genus *Sordaria*. *Neurospora Newsletter* 16: 14–17 (http://www.fgsc.net/sordnn16.html).

2   Glase, J. C. Tetrad and gene mapping in the fungus *Sordaria finicola*: http://www.wisc.edu/botit/img/Botany_130/Lab_Manual/kMeiosis.pdf.

3   The Fungal Genetics Stock Center: http://www.fgsc.net/.

Invertebrates

C. elegans

1   Anderson, P. (1995). Mutagenesis. *Methods Cell Biol* 48: 31–58.

2   Epstein, H. F. and Shakes, D. C. (eds.) (1995). *Caenorhabditis elegans*: modern biological analysis of an organism. *Methods Cell Biol* 48: 1–654.

3   Riddle, D. L., Blumenthal, T., Meyer, B. J., and Priess, J. R. (1997). *C. elegans II*. Cold Spring Harbor Laboratory Press, Cold Spring Harbor, NY. Available online at: http://www.ncbi.nlm.nih.gov/books/bv.fcgi?call=bv.View..ShowSection&rid=ce2.

4   Wormbase: http://www.wormbase.org/.

D. melanogaster

1   Ashburner, M. (1989). In: *Drosophila: A Laboratory Handbook*. Cold Spring Harbor Laboratory Press, Cold Spring Harbor, NY.

2   Flybase: http://flybase.bio.indiana.edu/.

3   Greenspan, R. (1997). *Fly-Pushing*. Cold Spring Harbor Laboratory Press, Cold Spring Harbor, NY.

4   Lindsley, D. L. and Grell, E. H. (1968). *Genetic Variations of Drosophila melanogaster*. Carnegie Institute Publication 624.

5   Lindsley, D. L. and Zimm, G. C. (1992). The Genome of *Drosophila melanogaster*. Academic Press, New York.

6   Wolfner, M. F. and Goldberg, M. L. (1994). Harnessing the power of Drosophila genetics. *Methods Cell Biol* 44: 33–80.

7   St Johnston, D. (2002). The cut and design of genetic screens: *Drosophila melanogaster*. *Nat Rev Genet* 3: 176–88.

8   Adams, M. D. and Sekelsky, J. J. (2002). From sequence to phenotype: reverse genetics in *Drosophila melanogaster*. *Nat Rev Genet* 3: 189–98.

Anopheles

1   Severson, D. W., Brown, S. E., and Knudson, D. L. (2001). Genetic and physical mapping in mosquitoes: molecular approaches. *Ann Rev Entomol* 46: 183–219.

Vertebrates

Zebrafish

1   Patton, E. E. and Zon, L. I. (2001). The art and design of genetic screens: zebrafish. *Nat Rev Genet* 2(12): 956–66.

2   The zebrafish website: http://zfin.org/.

Goldfish

1   Smartt, J. (1996). *Goldfish Varieties and Genetics*. Blackwell Science.

M. musculus

1   Justice, M. J., Noveroske, J. K., Weber, J. S., Zheng, B., and Bradley, A. (1999). Mouse ENU mutagenesis. *Human Mol Genet* 8(10): 1955–63.

Box 2.5   (*cont'd*)

2 Justice, M. J. (2000). Capitalizing on large scale mouse mutagenesis screens. *Nat Rev Genet* 1: 109–15.

3 Stanford, W. L., Cohn, J. B., and Cordes, S. P. (2001). Mouse genomic technologies: gene-trap mutagenesis: past, present and beyond. *Nat Rev Genet* 2: 756–68.

4 Copeland, N. G., Jenkins, N. A., and Court, D. L. (2001). Recombineering: a powerful new tool for mouse functional genomics. *Nat Rev Genet* 2: 769–79.

5 Yu, Y. and Bradley, A. (2001). Engineering chromosomal rearrangements in mice. *Nat Rev Genet* 2: 780–90.

Medaka

1 Wittbrodt, J., Shima, A., and Schartl, M. (2002). Medaka – a model organism from the far east. *Nat Rev Genet* 3: 53–6.

Rattus rattus

1 Jacob, H. J. and Kwitek, A. E. (2002). Rat genetics: attaching physiology and pharmacology to the genome. *Nat Rev Genet* 3: 33–452.

2 NIH Rat Genomics and Genetics website: http://www.nih.gov/science/models/rat/resources/index.html.

Homo sapiens

1 Collins, F. S. (1995). Positional cloning moves from perditional to traditional. *Nat Genet* 9(4): 347–50.

2 Ghosh, S. and Collins, F. S. (1996). The geneticist's approach to complex disease. *Ann Rev Med* 47: 333–53.

3 International Human Genome Sequencing Consortium (2001). Initial sequencing and analysis of the human genome. *Nature* 409: 860–921.

4 *Mendelian Inheritance in Man*. Available online at: http://www.ncbi.nlm.nih.gov/entrez/query.fcgi?db=OMIM&cmd=Limits.

5 Ott, J. (1999). *Analysis of Human Genetic Linkage*, 3rd edn. Johns Hopkins University Press, Baltimore, MD.

6 Scriver, C. R., Beaudet, A. L., Sly, W. S., Valle, D., Childs, B., and Kinzler, K. W. (2001). *Ber Vogelstein. The Metabolic and Molecular Bases of Inherited Disease*, 8th edn. Vols. I–IV. McGraw-Hill, New York.

7 The draft version of the human genome sequence with annotation of single nucleotide polymorphisms throughout the genome: http://www.ncbi.nlm.nih.gov/genome/guide/human/.

8 Venter, J. C. et al. (2001). The sequence of the human genome. *Science*: 1304–51.

9 Weiss, K. M. (1995). *Genetic Variation and Human Disease Principles and Evolutionary Approaches*. Cambridge University Press, Cambridge, UK.

Tyrannosaurus rex

1 Crichton, M. (1999). *Jurassic Park*. Ballantine Books, New York.

Plants

A. thaliana

1 Arabidopsis Information Resource at CSHL: http://stein.cshl.org/atir/biology/genome/.

2 Arabidopsis Information Resource at Stanford: http://www.arabidopsis.org/home.html.

3 Mutant Genes of Arabidopsis: http://mutant.lse.okstate.edu/genepage/genepage/html.

4 Meyerowitz, E. M. and Somerville, C. R. (eds.) (1994). *Arabidopsis*. Cold Spring Harbor Monograph 27, 1300 pp.

5 Weigel, D. and Glazebrook, J. (2002). *Arabidopsis: A Laboratory Manual*. Cold Spring Harbor Laboratory Press, Cold Spring Harbor, NY.

6 Page, D. R. and Grossniklaus, U. (2002). The art and design of genetic screens. *Nat Rev Genet* 3: 124.

Maize

1 Neuffer, M. G., Coe, E. H., and Wessler, S. R. (1997). *Mutants of Maize*. Cold Spring Harbor Laboratory Press, Cold Spring Harbor, NY.

2 The maize database: http://zmdb.iastate.edu/.

Wheat

1 The Wheat Genetics Resource Center: http://www.ksu.edu/wgrc/.

The complementation test

3

If the mutant hunts explained in the last chapter are done correctly and with real vigor, you now possess a large number of newly isolated mutants. The question becomes: how many different genes are represented in this collection of mutants? There have been cases where all of the newly recovered mutants were the result of separate mutational events in the same gene. However, there have also been cases where all of the newly recovered mutants identified new and separate genes. The most effective way to determine how many of your new mutations occurred in the same gene is to use the complementation test. A rigorous description of this test is provided in box 3.1, but we begin by discussing the complementation test in its simplest form.

3.1 The essence of the complementation test

Consider two independently isolated mutations, *m1* and *m2*. Mutation *m1* and mutation *m2* are both fully recessive, and *m1/m1* and *m2/m2* homozygotes produce similar mutant phenotypes. If *m1* and *m2* are in the same gene, double heterozygotes (*m1/m2*) will possess only mutant copies of the gene, and thus produce only the mutant phenotype. If the mutant phenotype is observed, we say that *m1* and *m2* "fail to complement each other" and that they define the same gene.

Suppose, however, that *m1* and *m2* are not in the same gene. These double heterozygotes have a wildtype phenotype because they carry one wildtype allele of both the m1 and m2 gene. In this instance, the two mutants are said to "complement each other" and to define different genes.

Formal definitions of the complementation test gain greater meaning in the context of examples. (An example of the complementation test as used in yeast is provided in box 3.2.) We begin here with an application of the complementation test in the fruit fly *D. melanogaster*. Consider the case of two strains (strain 1 and strain 2) of recessive wingless mutants, both of which lie in the same gene (a gene that is required for wing formation). We will denote the two mutations **a1** and **a2**; thus strain 1 flies are **a1/a1**, strain 2 flies are **a2/a2**, and wildtype flies are **A/A**. If we cross the **a1/a1** males to **a2/a2** females (or vice versa), we will always get **a1/a2** progeny, and all flies will have short wings. The crucial point is that

Box 3.1   A more rigorous definition of the complementation test

A more formal version of this test is the so-called *cis–trans* test. Using this test, two mutants are said to define the same functional unit (cistron) or gene *"if a heterozygote having (these) two different allelic recessives (arisen by independent mutations), one on one chromosome and the other on the homologous chromosome (trans or repulsion arrangement), has a recessive phenotype, or a more nearly recessive phenotype than the corresponding double heterozygote with both recessives on one chromosome and a normal homolog (cis or coupling arrangement),"* i.e.:

m1/+;+/m2 has a more nearly recessive phenotype than m1/+;m2/+

(Pontecorvo 1958, p. 37).

Buried in Pontecorvo's formalisms are three critical components of a successful application of the complementation, or *cis–trans*, test. First, Pontecorvo suggests that the test is applicable only to allelic mutants. (Pontecorvo's definition of allelism here is based on a segregational assay. He intended the test to be used only for mutations at similar sites on homologous chromosomes.) Second, Pontecorvo requires that the test be applied only to **recessive** mutations. Third, the so-called *cis* component of the test is critical because it rules out a very common reason for a misleading result from the complementation test, namely combined haplo-insufficiency. Combined haplo-insufficiency is a case where a combined reduction in the dosage of two different genes can produce a mutant phenotype.

because there are two different mutant alleles (**a1** and **a2**) in this genome, no normal **A** product can be produced.

Now suppose that the two mutations are located in different genes, and the normal products of both genes (**B** and **C**) are required to build a wing of normal length. Strain 3 is homozygous for a recessive loss-of-function mutation (**b**) in gene **B**, but carries two normal alleles of gene **C** (**bbCC**). Similarly, strain 4 is homozygous for a recessive loss-of-function mutation (**c**) in gene **C**, but carries two normal alleles of gene **B** (**BBcc**). Both **bb** and **cc** homozygotes produce short wings. If strain 3 males are crossed to strain 4 females, the resulting progeny are the genotype **BbCc**; they are doubly heterozygous at both genes. These progeny carry one wildtype (or normal functioning) allele of each gene (**B** and **C**). In this case, the wings of these flies have normal length because the wildtype alleles of the two genes provide normal copies of each protein. [*In this sense the transformation rescue experiments that are now possible in virtually every system (see box 3.3) may be considered to be permutations of complementation tests. If the introduced construct can "rescue" the function ablated by the mutant allele(s), then DNA carried by that construct defines the same gene.*]

The complementation test is only a test of gene function, and provides no information regarding the nature or position of the mutants. Two mutants that alter the same base pair in the same gene will fail to complement, just as will two very different mutants in the same gene. To determine whether or not two

Box 3.2   An example of using the complementation test in yeast

The baker's yeast *S. cerevisiae* can live as either a haploid or a diploid. Haploids can be mated to produce diploids, and the resulting diploids can be induced to undergo meiosis producing four haploid spores. This latter process is called sporulation. Haploid spores can be grown on Petri dishes to form haploid colonies. Diploid cells can produce diploid colonies.

Imagine that you have two newly induced mutants. The first mutant destroys a gene required for the cell to make the amino acid leucine. These cells cannot grow unless leucine is added to the media. This mutation is referred to as a leucine-minus auxotroph and is denoted as (*leu*–). The other mutant is a tryptophan-minus (*tryp*–) auxotroph (i.e. this mutation inactivates a gene whose protein product is required for the cell to synthesize tryptophan). Both of these mutations are recessive, therefore a diploid carrying one mutant and one normal copy of either gene can still synthesize the necessary amino acid. Both types of haploids will die if plated separately on minimal media containing neither leucine nor tryptophan. However, if the two haploids are mated to form a diploid, the *leu*– cell brings a functional tryptophan synthesis gene (denoted TRYP+) to this union, and the *tryp*– haploid brings a functional leucine synthesis gene (LEU+). The genotype of the resulting diploid is *TRYP+/tryp*–;*LEU+/leu*–. This diploid will be able to survive on minimal media because it carries functional genes for both leucine and tryptophan biosynthesis. In this sense, the two mutant-bearing genomes complement each other.

Now suppose you had two independently arising mutants that were both *leu*–. Leucine biosynthesis requires a succession of enzymatic steps, therefore one could easily imagine that the two mutants define different genes, and thus different enzymatic defects. Consider the hypothetical pathway drawn below:

Substrate $\longrightarrow$ I $\longrightarrow$ II $\longrightarrow$ III $\longrightarrow$ leucine

Enzyme	A	B	C	D
Gene	LEU1	LEU2	LEU3	LEU4

If the first mutant (*leu-1*) is in the gene that encodes enzyme A and the second mutant (*leu-2*) is in the gene for enzyme C, then the genotype of a *leu-1/leu-2* diploid should in fact be rewritten as *LEU1/leu1-1;LEU3/leu3-2*. You can easily see that in such a diploid the genes exist to make all four necessary enzymes.

But now suppose you find a third mutant (*leu-3*) that also defines the LEU1 gene. The genotype of a diploid created by mating a *leu-1* haploid and a *leu-3* haploid is best rewritten as *leu1-1/leu1-3*. There is no functional copy of the LEU1 gene. This diploid will remain a leucine auxotroph.

Thus, by creating diploids for various pairwise combinations of a large number of independently isolated leucine auxotrophs, one can quickly identify those mutants that define the same gene, or complementation group. You should realize that sporulating diploids with non-complementing mutants should only rarely produce LEU+ haploid spores, while such wildtype spores should be common when complementing diploids are sporulated. Why? How might the rare wildtypes derived from the non-complementing diploids have arisen? (The answer is presented in chapter 5.)

non-complementing mutants occur at the same or different sites in a given gene, one needs to turn to intragenic recombination. (The confounding of these two concepts, complementation and intragenic recombination, has a noble history in genetics, none the less, it should not be perpetuated.)

Box 3.3   Transformation rescue is a variant of the complementation test

In the matter of gene finding, *transformation rescue* is considered to be the gold standard of proof. If you can show that a homozygote for the mutant of interest can be "rescued" by the addition of a wildtype copy of your candidate gene, then you have cloned the gene defined by that mutant. If you are careful you will double check this conclusion by sequencing that gene in lines homozygous for one or more mutant alleles. In reality, the transformation rescue test is really just a permutation of the complementation test. Can a wildtype allele rescue the function of the mutant allele?

Could the transformation rescue test ever lie? It doesn't happen very often, but it does happen. The simplest and by far the most common error is a consequence of a phenomenon called "**multicopy suppression.**" Multicopy suppression is described in chapter 4 (section 4.8.1) in detail. For our purposes here, we need only note that if the rescue is done in yeast using a high copy number plasmid, then increasing the dosage of gene X can sometimes suppress a mutation in gene Y, perhaps by allowing the cell to bypass the defect. *Lesson:* never do transformation rescue using a multicopy

plasmid. Or if you do, first sequence the supposedly mutant alleles.

But even single copy rescue tests can still get you into trouble. Beall and Rio (1996) reported that mutants in the Drosophila *mus309* gene were rescued by a single copy insertion of a gene encoding the DNA repair protein Ku. Indeed, the Ku gene had been mapped to a small region that contained the *mus309* mutations. Unfortunately, it turns out that *mus309* mutations are not alleles of the Ku gene. Rather, they were shown to be alleles of the nearby *Dmblm* locus. *Dmblm* is a homolog of the human Bloom syndrome gene, which encodes a helicase of the RECQ family (Kusano et al. 2001a,b). It just turned out that both genes function in the repair of double strand breaks and that increasing the dose of Ku, even by 50%, can suppress the effects of homozygosity for loss-of-function mutations at the Ku gene. This misunderstanding, and perhaps some embarrassment, could have been avoided had the original workers sequenced the Ku gene in *mus309* homozygotes. Be careful out there: sometimes Nature has a nasty little mean streak.

3.2   Rules for using the complementation test

[handwritten: FULLY recessive mUTANT]

1. *The complementation test can be done ONLY when both mutants are fully recessive.* Because the complementation test works by revealing the presence or absence of a normal allele, it will work only for recessive mutants. (We present a method in box 3.4 for determining whether or not a given dominant mutation is allelic to one or more recessive alleles of a given gene.)

[handwritten: not need to have the same phenotype since it recessive it will show the weaker phenotype]

2. *The complementation test does not require that the two mutants have the same phenotype.* As we have noted previously, different mutations in the same gene can produce rather different phenotypes. Sometimes a mutant that alters, but does not destroy, function will have a weaker effect on the organism's phenotype than does a null or "knock-out" mutation. In these cases the double heterozygote usually exhibits the phenotype characteristic of the weaker of the two alleles. As an example, consider two different, independently isolated, mouse mutants. When homozygous in females, the first mutant, denoted *fs(a)*, results in sterility. The second mutant, denoted *l(b)*, is a recessive lethal.

Box 3.4   One method for determining whether or not a dominant mutation is an allele of a given gene, or how to make dominants into recessives by pseudo-reversion

If you need to determine whether or not a given dominant mutation is allelic to an existing tightly linked recessive, or to another closely linked and phenotypically similar dominant, your only hope is to induce "revertants" of the dominant allele. If you are lucky, some of these "revertants" won't be true revertants at all, but rather new loss-of-function alleles exhibiting a phenotype when homozygous. Such mutants are referred to as pseudo-revertants.

This approach depends on the dominant allele in question being a "gain-of-function" mutation, either an antimorph or a neomorph. The basic concept is simple: if the mutation is dominant because the gene now produces a poisonous product or produces the correct product at the wrong place or time, then the addition of a second "knock-out mutation" which ablates the ability of that allele to make *any* product should destroy the dominance of that allele. Indeed, the result should be a loss-of-function allele. Such mutants are referred to as "psuedo-revertants" because, although they appear to revert the "dominant" effects of the mutation, they are not true reversions of the original dominant mutation to wildtype. They are simply second mutational events in the same gene that inactivate the poisonous or misexpressed allele.

Recall the case of the *nod* gene in Drosophila discussed in chapter 1. The *nod* gene is required for female meiosis in Drosophila, but it is expressed in virtually all mitotically dividing cells. There are a number of recessive loss-of-function *nod* alleles that disrupt meiotic chromosome segregation when homozygous. There is, however, also a mutation that was originally called *l(1)TW-6cs* and mapped very close to the *nod* gene. This mutation exhibited a dominant meiotic phenotype that was identical to the phenotype exhibited by the loss-of-function *nod* mutations.

Rasooly et al. (1991) obtained three pseudo-revertants of *l(1)TW-6cs*, all of which no longer caused a meiotic defect when heterozygous. However, homozygotes for these pseudo-revertants displayed a meiotic phenotype identical to that observed for loss-of-function alleles. A series of complementation tests using the pseudo-revertants and existing loss-of-function alleles of *nod* revealed that all three of the pseudo-revertants failed to complement existing *nod* alleles or each other. By this test it was concluded that *l(1)TW-6cs* was in fact a dominant allele of nod and was renamed *nod^{DTW}*. Thus, in this case, allelism of a dominant antimorphic mutation could be demonstrated only by reverting the dominant to a recessive loss-of-function mutation. The point is, you can do a complementation test only with recessive mutations.

The reversion of dominant anti- or neomorphic alleles is well documented in the literature. Indeed, an excellent example of the reversion of a dominant allele of *nanos* (Arrizabalaga & Lehmann 1999) was presented in chapter 2. Other examples involve the reversion of dominant mutations at the *doublesex* (*dsx*) locus (Denell 1972; Duncan & Kaufman 1975; Belote et al. 1985), the *Sex lethal* (*Sxl*) locus (Cline 1978, 1984), the *Antennapedia* gene (Hazelrigg & Kaufman 1983), *Dichaete* (Russell 2000), *Enhancer of Split* (Nagel et al. 1999), and *Aberrant X segregation* (*Axs*) (Whyte et al. 1993).

These two mutants are tightly linked, as evidenced by tight linkage of both *fs(a)* and *l(b)* to some other mutation. The double heterozygote is sterile, exhibiting the weaker of the two phenotypes.

The finding of sterility in *fs(a)/l(b)* double heterozygotes suggests that the two mutations define the same gene; they fail to complement each other with respect to sterility. One explanation for this observation is that the *fs(a)* mutation is a hypomorphic allele, while the *l(b)* mutation is a nullomorphic allele. According to this explanation, the quantity of protein required for viability is less than the quantity required for fertility. As long as some protein is produced, the individual will survive, but a normal or near normal level of the protein is required for fertility. A second explanation argues that the wildtype protein has both a general function required for viability and a second, rather more specialized, function for fertility. In the case of a lethal mutation, the gene is disrupted to produce a null allele, while the *fs(a)* mutation, which is a recessive female sterile, specifically disrupts the site on the protein required for fertility.

3 *There are cases where the phenotype of a double heterozygote is more extreme than that of either homozygote.* There are several well-described instances in which the phenotype of a double heterozygote for two mutants within a given gene is considerably more extreme than that of either of the two homozygotes. This phenomenon is referred to as "negative complementation" (Fincham 1966). Negative complementation presumably reflects the ability of the two abnormal protein products to form a dimer or multimer that is not only non-functional (as are homodimers of the two mutant proteins), but poisonous as well. Although this phenomenon is best described in Drosophila (Raz et al. 1991; Bickel et al. 1997), it has also been described in several other instances (Fincham 1966).

3.3   How might the complementation test lie to you?

Although the complementation test is generally an excellent tool, there are well-documented examples of the test yielding erroneous results. Some of these are discussed below.

The first example describes cases in which two mutations in the same gene complement each other (**intragenic complementation**). The fact that some proteins have different functional domains, which can be separately mutated, raises an intriguing question. Suppose a given gene encoded a protein with several separate functional domains, and that mutants in a given domain sometimes only inactivated that specific function. Could two mutations exist in the same gene, but map in separate functional domains, and thus complement each other? There is a gene in Drosophila called *rudimentary* that encodes a protein exhibiting three spatially and functionally distinct domains. Missense mutants that alter the first domain of this gene can be complemented by mutants that specifically alter the sequence of the second or third domain. All that the cell, or fly, seems to require is that at least one functional copy of all three domains is present in the cell. [The difficulties inherent in this analysis of *rudimentary* will be discussed in detail in chapter 6 (box 6.1) – suffice it to say that working out structure function relationships between genes and their protein products using intragenic complementation is a tricky business.]

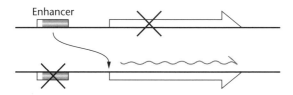

Figure 3.1 Transvection (an example of pairing-dependent complementation). Intragenic complementation in which the non-mutant enhancer functions in *trans* to activate transcription of the non-mutant gene on the homologous chromosome

Another example of intragenic complementation is provided by cases in which one of the two mutants lies in an upstream regulatory element, and the other lies within the coding region. In at least some organisms, including Drosophila, there is a sufficient degree of somatic pairing of homologous chromosomes to allow the functional copy of the regulatory region to act properly on the gene carried by the homologous chromosome. As shown in figure 3.1, this allows the functional regulatory element on the chromosome with the mutant in the coding region to "reach over" and activate the normal sequence promoter region on its homolog. This complementation is dependent on proper somatic chromosome pairing and is referred to as transvection (discussed in box 3.5). A detailed consideration of the pairing-dependent complementation patterns observed for the yellow and BX-C genes in Drosophila is found in chapter 6 (section 6.6).

Could two mutations in different genes ever fail to complement each other? The answer to this question is "yes." This phenomenon was first noted by Calvin Bridges in the 1930s, and was referred to as dominant enhancement or unlinked non-complementation. It is now referred to as second-site non-complementation (SSNC), and is sufficiently important to require its own section, which follows below.

3.4 Second-site non-complementation (SSNC) (non-allelic non-complementation)

In rare cases, two mutants in different genes, which by themselves are fully recessive, can create a mutant phenotype in double heterozygotes. In other words, although individuals heterozygous for either *m1* or *m2* alone (*m1/+* and *m2/+*) are wildtype, doubly heterozygous individuals (*m1/+;m2/+*) exhibit a mutant phenotype. This phenomenon is referred to as *non-allelic non-complementation* or *second-site non-complementation*, as described below. We divide this phenomenon into three types of SSNC. In Type I, SSNC mutant forms of the two different proteins interact to produce a poisonous product. In Type II SSNC the mutant form of one protein sequesters the wildtype form of the other protein into an inactive complex. In Type III the simultaneous reduction of the two proteins results in a phenotypic abnormality. Only Type I and Type II SSNC suggest physical interaction between the two proteins.

Box 3.5   Pairing-dependent complementation: transvection

There are many examples in Drosophila, in which intragenic complementation can be disrupted by heterozygosity for chromosomal rearrangements [for a review, see Wu and Morris (1999)]. These effects appear to reflect the ability of various enhancer-like elements to function in *trans* in an organism like Drosophila, which has ubiquitous somatic pairing. That is to say, an *m1/m2* heterozygote (in which *m1* defines an enhancement element and *m2* defines the coding sequence) may exert a wildtype appearance because the wildtype copy of the enhancer born by the *m2* chromosome can activate the wildtype coding sequence born by the *m1* chromosome (see figure 3.1). However, this *trans*-activation will be suppressed by genetic rearrangements that suppress pairing.

Transvection was first described and demonstrated at the *bithorax* complex (BX-C) in Drosophila by Ed Lewis (1954). Using above-ground nuclear detonations to produce rearrangements, Lewis demonstrated that even breakpoints located quite

far away on the same chromosomal arm from the gene in question could generate transvection. Homozygosity for those breakpoints does not usually cause transvection. Subsequently, transvection has been demonstrated at multiple loci in Drosophila [for a review, see Ashburner (1989) but cf. Gelbart (1982)]. Perhaps the most interesting of these cases involves the pairing-dependent interaction of the *zeste* and *white* genes. Homozygotes for loss-of-function mutations at the *zeste* locus can modify the expression of the *white* locus, but only when two copies of the *white* locus are paired (Green 1977; Jack & Judd 1979; Gelbart 1982). An elegant comparison of transvection at several loci in Drosophila is provided by the work of Smolik-Utlaut and Gelbart (1987), and a mechanistic basis for the observed differences is provided by Golic and Golic (1996). There is some evidence that transvection may also occur in Neurospora (Perkins 1997). Indeed, similar effects have been noted in a variety of organisms (Tartof & Henikoff 1991; Matzke et al. 2001).

3.4.1   Type 1 SSNC (poisonous interactions): the interaction is allele-specific at both loci

This type of second-site non-complementation is both the rarest and the most interesting. By definition it requires allele-specificity at both genes, and neither allele can be a null mutant. In this sense, the double mutant combination may be thought of as a "synthetic antimorph" because the combination of the two mutant proteins produces a poisonous gene product. This type of interaction is often explained by asserting that the two mutant proteins physically interact to produce a poisonous protein dimer or complex (figure 3.2). However, the proof of such a hypothesis requires biochemical evidence for the physical association of the two mutant proteins and elucidation of the mechanism by which they create a poisonous effect. Moreover, as our second example will show, there are cases where this type of highly specific genetic interaction is not mediated by the direct physical interaction of the two gene products.

An example of Type 1 SSNC involving the alpha- and beta-tubulin genes in yeast

There are two genes in yeast for alpha-tubulin and one for beta-tubulin. The *TUB1* and *TUB3* genes encode variants of alpha-tubulin, while the *TUB2* gene

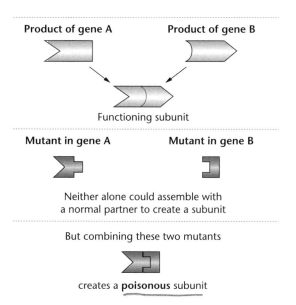

Figure 3.2 One way of imaging how two different mutant protein subunits might interact to create a poisonous dimer molecule

encodes beta-tubulin. In order to identify genes whose products interacted with beta-tubulin, Tim Stearns and David Botstein (1988) screened for mutants that displayed cold-sensitive SSNC in the presence of a recessive cold-sensitive allele of beta-tubulin (*tub2cs*). This screen is diagrammed in figure 3.3. The focus on cold-sensitivity here reflects a prejudice arising from phage genetics that cold-sensitivity is a hallmark of mutations that affect protein assembly or protein–protein interactions.

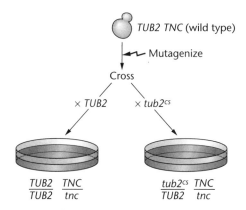

Figure 3.3 A screen for second-site non-complementing mutants involving the tubulin genes in yeast. *TUB2* is the gene for beta-tubulin, *TNC* is a tubulin non-complementer gene. A mutant that forms a Cs⁻ diploid with a *tub2cs* mutant, but a Cs⁺ diploid with wildtype, is a non-complementer. The mutation responsible for the non-complementation is then tested for linkage to the *TUB2* locus. [An excellent summary of the basic techniques of yeast mutant hunting can be found in Forsburg (2001)]

Stearns and Botstein created a large collection of mutagenized haploid yeast colonies. They grew each colony from a single mutagenized cell, and required that these cells be viable at the permissive temperature. Each of these lines was then separately mated to a haploid strain carrying the recessive *tub2cs* mutant. Thus, the resulting diploids were heterozygous for both the new mutants from the mutagenized haploid line and the recessive *tub2cs* mutant. Diploids in which the two mutations show second-site non-complementation should produce the cold-sensitive phenotype. In the actual experiment, cells from each of the haploid colonies were mated to haploids carrying the recessive *tub2cs* mutant, and other cells from the same colonies were mated to wildtype haploids. All of the resulting diploids were then plated at the restrictive (non-permissive) temperature.

There are four types of new mutations that could cause the resulting cells to fail to grow at the restrictive temperature. These include mutants that block mating (diploid formation), new dominant cold-sensitive mutations, new mutants in the TUB2 gene, and lastly the sought after second-site non-complementers. Fortunately, the first two types of mutants are easily recognized because they fail to produce colonies when mated to wildtype cells. The last two classes can be distinguished by a simple segregational assay.

The 20,000 colonies grown from mutagenized cells were screened for failure to complement either of two *tub2cs* alleles (*tub2cs-104* and *tub2cs-401*), and only three non-complementers were found. Two of these non-complementers turned out to be new alleles of TUB2. However, one of these mutants, designated *tnc1*, turned out to be a mutation at another gene that failed to complement the *tub2cs* gene. Thus *tnc1/+ tub2cs/+* double heterozygotes show the same cold-sensitive lethality that is observed in tub2cs homozygotes.

The *tnc1* mutant fails to complement *tub2cs-401* at 16°C, but weakly complements *tub2cs-104* at the same temperature. That is, the interaction of *tnc1* with the *TUB2* gene is allele-specific. Some allelic combinations were strongly cold-sensitive, others weakly cold-sensitive, and still others were not at all cold-sensitive [see Table 6 in Stearns and Botstein (1988)]. The *tnc1* mutant is also cold-sensitive by itself, and exhibits a defect in spindle assembly even at the permissive temperature. Both the cold-sensitivity of *tnc1* and its inability to complement *tub2cs-401* segregate as simple single gene traits, and the two traits are not separable by recombination.

Perhaps not surprisingly, *tnc1* turned out to be an allele of TUB1, the gene for alpha-tubulin. (The mutation was then renamed *tub1-1*.) This was an important finding because, prior to this study, no viable mutant alleles of the *TUB1* gene were available. (Null mutants at this gene, which can be created by various techniques, are lethal in haploids. Indeed, even as heterozygotes, such mutants have dominant effects on viability, and the diploid heterozygotes rapidly become aneuploid.) Moreover, Stearns and Botstein continued their efforts by screening for second-site non-complementers of *tub1-1* (*tnc1*). This screen yielded both a new cold-sensitive allele of TUB2 (*tub2-501*), and more critically the first recovered point mutation in the TUB3 gene (*tub3-1*). The only previous existing alleles of TUB3, the second gene encoding alpha-tubulin, were null mutations generated by gene disruption.

It is curious though that both the point mutant and the null allele of TUB3 failed to complement the *tub1-1* allele. To quote Stearns and Botstein, "this means that the failure to complement in this case is not due to a specific protein–protein interaction. Instead, it seems that a reduction in the total level of functional alpha-tubulin in the *TUB1/tub1*; *TUB3/tub3* double heterozygote causes the failure to grow at 14°." Indeed, this case of "combined haploinsufficiency" is a model for the third type of SSNC which will be discussed later in this chapter.

The Stearns and Botstein (1988) paper is widely viewed as the hallmark on SSNC research. First, it was the first such report to appear and the first concrete example that such screens can work. The screen did identify a mutant in a gene (*TUB1*), whose product, alpha-tubulin, physically interacted with beta-tubulin and provided the first mutant in the second alpha-tubulin gene (*TUB3*). (Mutants in *TUB1* and *TUB3* did not exist prior to this screen.) Second, the study was truly elegant. Unfortunately, it would be one of very few such screens to be successful. We now consider a similar screen for actin non-complementers in yeast that was less successful in identifying interacting proteins.

An example of Type 1 SSNC involving the actin genes in yeast

Following the studies of Stearns and Botstein, David Drubin and collaborators set out to determine whether or not a screen for second-site non-complementers in yeast could be used to identify genes whose protein products interacted with actin (Vinh et al. 1993; Welch et al. 1993). These studies followed an extensive analysis by Drubin of mutants that can suppress a temperature-sensitive allele of actin (*act1-1*). That study is discussed in the following chapter (section 4.7.1).

Drubin and collaborators began with the observation that null mutants in two genes known to encode actin-binding protein (*SAC6* and *ABP1*), failed to complement the temperature-sensitive phenotype caused by a mutation in the *ACT1* gene (figure 3.4). This observation emboldened them to attempt to identify novel genes whose protein products interact with actin, by screening for second-site mutants that fail to complement *act1-1* or *act1-4*, two temperature-sensitive alleles of *ACT1*. The screening of more than 55,000 mutagenized colonies yielded a total of 14 extragenic non-complementing mutants and 12 new alleles of *ACT1*.

The 14 SSNC mutations they isolated were shown to be alleles of at least four different genes, *ANC1*, *ANC2*, *ANC3*, and *ANC4* (actin-non-complementing). These mutations exhibited several properties that made it likely that they defined genes of interest, i.e. genes that encode actin-interacting proteins. First, mutations in the ANC1 gene were shown to cause defects in actin organization. Indeed, the phenotypes observed in the presence of *anc1* mutants alone were similar to those caused by *act1* mutations [see also Welch and Drubin

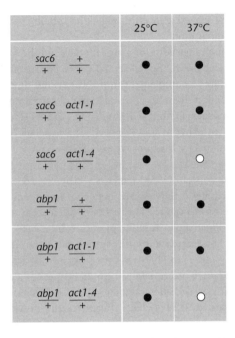

		25°C	37°C
$\dfrac{sac6}{+}$ $\dfrac{+}{+}$		●	●
$\dfrac{sac6}{+}$ $\dfrac{act1\text{-}1}{+}$		●	●
$\dfrac{sac6}{+}$ $\dfrac{act1\text{-}4}{+}$		●	○
$\dfrac{abp1}{+}$ $\dfrac{+}{+}$		●	●
$\dfrac{abp1}{+}$ $\dfrac{act1\text{-}1}{+}$		●	●
$\dfrac{abp1}{+}$ $\dfrac{act1\text{-}4}{+}$		●	○

Figure 3.4 Mutants in genes whose protein products interact with actin (*sac6* and *abp1*) fail to complement the temperature-sensitive *act1-4* mutation but fully complement the temperature-sensitive *act1-1* mutant at 37°C (i.e. they show actin allele-specific non-complementation). Filled circles indicate growth of the genotype on petri dishes incubated at the indicated temperatures. Open circles indicate little or no growth. (Figure courtesy of Diana Hiebert.)

(1994)]. Second, the observed non-complementation was allele-specific. Third, when mutant alleles of four ANC genes (*ANC1*, *ANC2*, *ANC3*, and *ANC4*) were tested for genetic interactions with null alleles of known actin-binding protein genes, an *anc1* mutant allele failed to complement null alleles of the *SAC6* and *TPM1* (genes that encode the yeast actin-binding proteins fimbrin and tropomyosin, respectively). Fourth, synthetic lethality between *anc3* and *sac6* mutants, and between *anc4* and *tpm1* mutants, was observed. These rather complex genetic interactions are displayed in figure 3.5. (Synthetic lethality is another tool for identifying genes whose products might interact either physically, or in the same pathway or process. A discussion of this technique is found in box 3.6.)

Unfortunately, things didn't work out that way. The Anc1p protein turns out not to be a physical partner of actin. Rather, Anc1p is a transcription factor that is also known as Tfg3 or TAF30 (Cairns et al. 1996). Specifically, it is a yeast-specific subunit of the transcription factor TFIIF. Indeed, the Anc1 protein is a component of the so-called "mediator complex," whose interaction with the carboxy-terminal repeat domain of RNA polymerase II enables transcriptional activation. Deletion of ANC1/TFG3 results in diminished transcription. Thus mutants in the ANC1 gene appear to interact with the act1 mutants not by physical interaction of mutant gene products, but instead by global effects on the transcription process. The nature of that interaction has not been elucidated; nor is it likely to be.

This actin SSNC example points to a caveat in the use of non-complementation screens to find interacting proteins: many of the ways in which a cell might create a given defect will not involve a physical interaction with the mutant protein of interest, and thus may not be of real interest to you. And no matter how clever your secondary screens are, you still might get mutants in genes you weren't looking for or interested in.

Box 3.6   Synthetic lethality and genetic buffering

As first noted by Bender and Pringle (1991), one technique for identifying genes whose products function in the related or parallel pathways is to look for cases of synthetic lethality. Following Hartman et al. (2001), "mutations in two different genes are said to be synthetically lethal if either mutation is viable in an otherwise wildtype background, but the combination of both alleles prevents growth." This definition works well in an organism like yeast, that can be grown in a haploid phase, but may become somewhat more awkward for diploid higher eukaryotes. In such cases, the synthetic lethality may define mutations that are recessive in such a way that a heterozygote for a loss-of-function at either gene is viable, but the double heterozygote is lethal. (This latter case is clearly an example of second-site non-complementation.) Alternatively, two dominant mutations may be synthetically lethal, or a recessive mutation at locus A may be synthetically lethal with some dominant allele of another gene even as a heterozygote. The simple test here is that if both alleles can survive on their own, but kill the organism in combination, they are said to be synthetically lethal. Again we quote Hartman (2001), "Synthetic lethality defines a relationship where

the presence of one gene (A) allows the organism to tolerate genetic variation (b) in another gene (B) that would be lethal in the absence of that first gene (A)."

We have discussed the use of semi-lethality to describe the relationships between the various actin non-complementing mutants. A more detailed exploration of interactions using semi-lethality involves the study of the genes whose proteins mediate the secretory pathway in yeast (figure B3.1). Synthetically lethal interactions have been found for many genes in this pathway that can be divided into ten discrete steps or components. Indeed, more than half these interactions involve genes whose products act at the same step (e.g. translocation to the Golgi Apparatus). Of 173 synthetic lethal interactions observed in this process: 116 involve the same component of the process; 68 involve proteins acting in different components of the process (a few genes act in two or more components); and 53 involve a gene that is not involved in secretion (Hartman et al. 2001).

Hartman et al. (2001) describe these interactions in terms of "buffering," i.e. the ability of one gene to buffer the effects of a loss-of-function mutation at another. The loss of the buffering

Box 3.6 *(cont'd)*

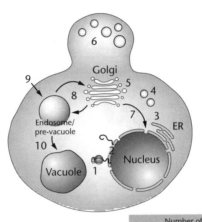

Step	Function	Number of genes	Number of synthetic lethal interactions		
			Same step	Different step	Not secretion
1	Translocation to Golgi	23	6	0	4
2	Maturation in ER	16	10	0	2
3	Vesicle budding	12	11	3	2
4	Vesicle fusion	18	11	13	1
5	Transport to Golgi	11	2	9	3
6	Fusion with plasma membrane	30	43	13	12
7	Retrieval	20	15	9	1
8	Vacuolar targeting	16	6	6	4
9	Endocytosis	20	12	12	22
10	Endosome to vacuole	7	0	3	2
	Total	173	116	68	53

Figure B3.1 The secretory pathway in yeast. (Adapted from Hartman et al. 2001)

gene reveals the defect in the second gene, resulting in lethality. They distinguish between "intrinsic buffering," in which both genes involved in a synthetic lethal interaction function in a single component or pathway, and "extrinsic buffering," in which the two genes involved in a synthetic lethal interaction function to produce proteins that act in different pathways. These authors speculate on the various mechanisms by which intrinsic buffering might function. These include the existence of various systems of feedback regulation that can optimize flow through a

biochemical process even in the presence of a serious deficiency at a single step. Extrinsic buffering interactions may reveal pathways that are functionally, if not biochemically, redundant. They also cite an example in which certain DNA repair pathways may be able to compensate for a defect in DNA replication. Things will be fine if the cell carries only the DNA replication defective mutation, but if one cripples the repair system as well by adding a second mutation, the cell will die.

Hartman et al. (2001) extended this study by examining all known synthetic lethal interactions

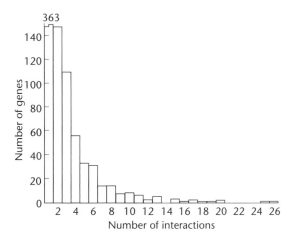

Figure B3.2 All known synthetic lethal interactions recorded in the yeast proteome database. For most genes, only one synthetic lethal relationship has been reported, and only a few genes have ten or more such interactions. (Adapted from Hartman et al. 2001)

that are recorded in the yeast proteome database (figure B3.2). These data reveal a startling specificity to synthetic lethal interactions. They found that a defect in a single gene was usually observed to be synthetically lethal with mutants in only one other gene, and few genes had more than ten interactions. These data are not comprehensive, but additional screens for mutations that are synthetically lethal with a given single mutation suggest that on average about three or four genes can mutate to synthetic lethality with a mutant allele of a given gene.

A thorough discussion of the execution of synthetic lethal screens in yeast may be found in Guarente (1993), Appling (1999), and Forsburg (2001).

The existence of synthetic lethal interactions thus provides a truly powerful tool for identifying two genes that act in the same component of a biological process, and in identifying redundant or parallel pathways.

The significance of this phenomenon is elegantly revealed by the work of Rutherford and Lindquist (1998). These workers reported that loss-of-function mutations in the Hsp90 chaper-

onin gene in Drosophila revealed cryptic mutations in genes that affected nearly every adult structure of the fly. These variations were masked by the wildtype allele of Hsp90 and were revealed when Hsp90 is mutant or pharmacologically impaired. When enriched by selection, these variants are able to express the abnormal phenotype in the absence of the Hsp90 mutation.

The observation that the effects of these cryptic mutations can be revealed by pharmacological agents is important because it provides a mechanism for understanding the inheritance of some apparently "acquired characters" [for a review, see McLaren (1999)]. This explanation provides a frame-work for understanding the work of Waddington in the mid-1900s, and his theories of canalization and genetic assimilation. A modern "proof" of the correctness of this explanation for Waddington's work is provided by the studies of Gibson and Hogness (1996). In these experiments, flies with defects in fly segment identity were selected as "phenocopies" following treatment with ether. Molecular analysis revealed that the ether vapor treatment simply unmasked what is normally silent genetic variation at the Drosophila *Ubx* gene.

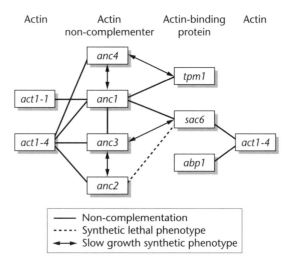

Figure 3.5 Multiple types of genetic interactions with actin mutants in yeast. Genetic interactions between *anc* mutations, between *anc* and *act1* mutations, and between *anc* and actin-binding protein mutations are illustrated. (Adapted from Welch et al. 1993)

3.4.2   Type 2 SSNC (sequestration): the interaction is allele-specific at one locus

In this second type of second-site non-complementation, we observe allele-specificity only for one of the two genes. The mutation at the other gene needs only to be a loss-of-function allele; indeed, a deficiency of the second gene will usually work just fine.

One useful way to conceive of this process at the molecular level has been advanced by Fuller (1986). Imagine a structural protein complex that is comprised of two protein subunits, A and B. In this case, heterozygosity for a null mutant of *B* (denoted *B/b*) has no phenotypic consequences by itself, because only half the quantity of the A–B heterodimer is necessary for the cell to function. Now imagine that the aberrant protein produced by a specific mutant allele at the *A* gene may have the capacity to sequester some quantity of the normal product of the *B* gene in a fashion that renders it inactive (figure 3.6). If a mutant form of A (A^*) is present that binds normally to a B subunit but then renders the complex inactive, then in an A/A^*; B/b double heterozygote, half of the available B protein will be bound up in A^*–B inactive complexes. As a consequence, the level of functional A–B complex will be only a quarter of the normal level. This

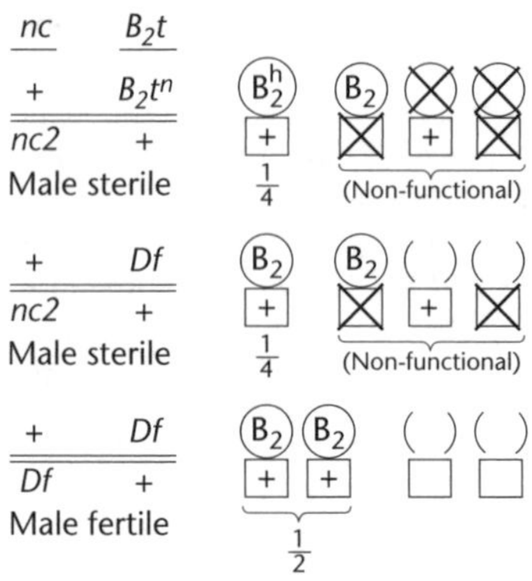

Figure 3.6 A null mutant at the B_2t gene (designated B_2t^n) fails to complement a missense mutant (*nc2*) as does a deficiency (*Df*) for the B_2t gene. However, a deficiency (*Df*) for the B_2t gene does complement a deficiency for the *nc* gene. Open circles or squares indicate functional (w.t.) proteins. Circles or squares with an "X" through them indicate mutant proteins. Broken symbols (e.g. ( ) or [ ]) indicate an absence of product of that gene. (After Fuller 1986)

difficulty is further compounded if the polyprotein is made up of two B subunits and two A subunits, and if a single A* subunit can poison the entire complex. It is important to realize that such an effect will not be created in an individual heterozygous for a null allele (*a* or *b*) at both loci. In this case, the available A protein and the available B protein will form half the normal level of functional complex, a level already known to be sufficient for a wildtype function. Because the second-site non-complementation observed here reflects the binding of a non-functional copy of one protein to normal copies of the other, this type of SSNC is characterized by allele-specificity at only one of the two genes (i.e. the one encoding the non-functional product or A*). The mutants at the other gene can be null alleles.

An example of Type 2 SSNC involving the tubulin genes in Drosophila

The yeast experiments performed by Stearns and Botstein (1988) were stimulated by previous work done in Drosophila. Fuller and her collaborators (Hays et al. 1989) recovered a mutation in an alpha-tubulin gene, denoted *nc33*, that failed to complement mutants in a beta-tubulin gene (*B2t*) in flies. The *B2t* gene encodes a variant form of beta-tubulin that is synthesized only during spermatogenesis.

As a homozygote, *nc33* behaves as a male sterile, but *nc33/+* males are quite fertile. Flies heterozygous for *nc33* and for various recessive alleles of *B2T*, including a functionally null (*B2t^n*), are sterile (Hays et al. 1989). In other words, although *nc33/+* or *B2t^n/+* flies are fertile, *nc33/+; B2t^n/+* flies are sterile (see figure 3.6). Thus, in a fashion similar to that described above, we see a phenotype resulting from double heterozygosity for recessive mutations at both the alpha- and beta-tubulin genes. The interaction of *nc33* with *B2t* mutations is not mimicked by a deficiency for the alpha-tubulin gene. The interaction requires a specific allele of the alpha-tubulin gene. Even in the presence of two normal alpha-tubulin alleles, the expression of the *nc33* mutant form of alpha-tubulin can still cause a sterile interaction with some *B2t* alleles, albeit at a reduced frequency.

But the critical difference between this interaction and those observed by Stearns and Botstein in yeast is that the most extreme phenotype is obtained with a null allele of the *B2T* gene. It appears that the *nc33* mutant alpha-tubulin protein acts to sequester the normal beta-tubulin subunits into inactive dimers. As a result, the total number of functional tubulin dimers is reduced below some critical threshold. Approximately half of normal levels, as would be observed in the presence of heterozygosity for the beta-tubulin null mutant alone, is apparently good enough. But reducing that concentration further, to a quarter of normal levels, impairs spermatogenesis.

Can the Fuller screen be generalized to look for other proteins that interact with alpha- and beta-tubulin? The studies described above suggested an exciting new method for identifying genes encoding other proteins that interact with alpha- and beta-tubulin, and such screens were performed by Fuller and her colleagues. Although these screens yielded a number of important and interesting new mutants [cf. Green et al. (1990); Regan and Fuller (1988)], it is not clear that the proteins encoded by these genes have any obvious functional relationship to alpha- and beta-tubulin. One of the best characterized of these second-site non-complementing genes is the *haywire* gene.

The product of the gene defined by the *haywire* mutant seemed to be an ideal candidate for a microtubule-interacting protein on several grounds. First, the interaction of *haywire* mutants with *B2t* null mutations is allele-specific and is not mimicked by a deficiency for the *haywire* gene (Regan & Fuller 1988). Second, the non-complementing allele of *haywire* (denoted *haync2*) is a recessive male sterile mutant in the presence of two normal alleles of *B2t*. Better yet, males homozygous for *haync2* have defects in the three major microtubule-based processes in which the testis-specific beta 2-tubulin participates, namely: meiosis, flagellar elongation, and nuclear shaping. These phenotypes were truly consistent with a role for the *haywire* gene product in general microtubule function, and a role in which the haywire protein directly interacted with microtubules.

Sadly, things turned out to be not so simple. When the *haywire* gene was cloned it was shown to encode a general RNA polymerase II transcription factor that is also essential for nucleotide excision repair (Mounkes et al. 1992; Mounkes & Fuller 1999). (Recall a very similar story for the yeast mutant *anc1* that was described in the previous section.) Presumably, the aberrant *haywire* gene product results in the impaired expression of some gene or set of genes in a fashion that cripples meiosis and spermatogenesis in a cell with reduced levels of beta-tubulin.

The lesson here is simple: by creating a male germline with greatly reduced levels of beta-tubulin subunits, one is creating a rather fragile state in which further reductions in perhaps a large number of proteins can result in dramatic defects in microtubule-based processes. It is not what Fuller sought to find, but it is exactly what she asked for. Her screen was not for genes whose protein products interacted with beta-tubulin, but rather for mutants that made a normally recessive *B2t* null mutation behave as a dominant.

We might also add that Calvin Bridges noted almost a century ago that mutants known as *Minutes* act as *dominant enhancers* of many recessive mutants in Drosophila. This is to say, double heterozygotes for the recessive mutant (**a**) and for the *Minute*, denoted *A/a*; *Minute/+*, often displayed the phenotype characteristic of homozygotes for the recessive mutant. Minute mutations turn out to define ribosomal protein genes in Drosophila, and as heterozygotes reduce the overall level of translation. In such heterozygotes the reduced level of wildtype gene product produced by the normal (**A**) allele is insufficient to produce a normal phenotype.

Thus a screen for second-site non-complementers can be considered to be a screen for dominant enhancers, especially in the case where the interaction is not allele-specific at both loci. Although screens for dominant enhancers can be powerful (see below), such a screen can also dredge up a host of undesired interacting loci. The maxim goes as follows: *Nature is both eminently fair and eminently cruel. Nature is fair because it will answer any question you ask, and cruel because Nature will answer the question you actually ask, not the question that you thought you were asking.*

But surely, some of the *B2t*-interacting genes must encode the types of microtubule-associated proteins that Fuller and her collaborators were seeking? To date, none of the interacting genes yet identified by Fuller and colleagues has been shown to encode a protein that physically interacts with microtubules.

However, the genetic analysis of the *whirligig* locus is encouraging (Green et al. 1990). The interaction of *whirligig* mutants with *B2t* null is allele-specific, and *whirligig* mutants have an interesting phenotype on their own. Two copies of the *whirligig* locus are necessary for male fertility. Both a deficiency of *whirligig* and loss-of-function alleles are dominant male sterile mutations, even in a genetic background wildtype for tubulin. This dominant male sterility is suppressed if the flies are also heterozygous for a null allele of *B2T*, for a deficiency of *alpha-tubulin*, or for the *haync2* allele. These results suggest that it is not the absolute level of *whirligig* gene product that matters, but rather its level relative to tubulin that is important for normal spermatogenesis. The phenotype of homozygous *whirligig* mutants suggests that the whirligig product plays a role in organizing the microtubules of the sperm flagellar axoneme. The flagellar axonemes show multiple anomalies in microtubule organization. It would certainly not surprise us if the *whirligig* gene product was eventually shown to encode exactly the type of protein that Fuller and her colleagues set out to find, namely a protein that physically interacts with tubulin.

An example of Type 2 SSNC in Drosophila that does not involve the tubulin genes

Dan Kiehart and his students subsequently applied SSNC to identify a group of genes whose protein products interact with non-muscle myosin in Drosophila (Halsell & Kiehart 1998; Halsell et al. 2000). In Drosophila, the *zipper* (*zip*) gene encodes the non-muscle myosin protein. Mutants in this gene produce malformed legs in the adult, a phenotype that is easy to assay. One can thus test for second-site non-complementation of the mutants in the *zip* gene by screening for mutants that behave as dominant enhancers of appropriate *zip* mutants. Although null alleles of *zipper* are not known to display second-site non-complementation, a missense allele (*zip[Ebr]*) proved useful as bait in a screen for "dosage-sensitive second-site non-complementers." Indeed, the *zip[Ebr]* allele was specifically chosen for this study because it had already been shown to display SSNC with certain alleles of a set of X chromosomal genes (the *broad complex* genes).

Kiehart and colleagues scanned through 70% of the Drosophila genome by testing a large collection of deficiencies for their ability to display second-site non-complementation with *zip[Ebr]*. They tested 158 such deficiency heterozygotes for their ability to produce malformed legs in flies that were simultaneously heterozygous for *zip[Ebr]*, and in doing so surveyed over 70% of the fly genome.

The technique that Halsell and Kiehart (1998) used is conceptually rather different from the studies we have seen above. Rather than creating new non-complementing mutants, they used an available collection of chromosomal deletions (each of which removes, on average, 50 to 100 genes). The question then becomes: which genes, or chromosome regions, strongly enhance the *zip[Ebr]* mutation when present in only one dose. (Remembering that all of these chromosomal regions are sufficient for a normal phenotype in *zip[Ebr]* heterozygotes when present in two doses.) The benefit of this technique is its relative ease. The majority of the genome can be scanned in less than 200 crosses. The potential weakness is that the interaction you seek has to be dosage-sensitive; one is not testing mutant against mutant, but rather asking for which genes does reducing the dosage by one copy become a problem in the presence of the *zip[Ebr]* mutation. (If you think about it, this is really the reverse of screening for point mutants that interact with a null mutant in the *B2t* gene.)

Halsell and Kiehart identified two chromosomal deficiencies that strongly enhanced the phenotype of the *zip[Ebr]* mutant in flies heterozygous for the *zip[Ebr]* mutant (i.e. the *zip[Ebr]*/+; *Df*/+ double heterozygotes displayed an obvious malformed leg defect) and 17 weekly interacting loci. From among these 19 or so deficiencies, they have been able to identify three whose interaction with zipper can be explained by individual genes located in these deficiencies. These include genes encoding cytoplasmic myosin, collagen IV, and the signal transduction protein RhoA. In the cases of cytoplasmic myosin and collagen IV, the identity of the interacting loci was determined by testing overlapping deficiencies and by testing known mutants at these loci for their ability to interact with *zipper*.

The route to the discovery of the RhoA interaction was rather more complex and requires a more detailed discussion. This gene was initially identified by two independently isolated EMS-induced recessive lethal mutants (known as *E3.10* and *J3.8*) that mapped within one of the interacting deficiencies. Both of these mutants failed to complement the *zip[Ebr]* mutant, and failed to complement each other with respect to recessive lethality. However, it was not at all clear from these studies which gene these mutants defined.

The answer came by the serendipitous finding of another, but related, interacting gene. Halsell et al. (2000) had performed a second screen for second-site non-complementation by testing 268 different transposon (P element) insertions for their ability to interact with *zipper*. In each case these insertions produced recessive lethal mutants (i.e. insertion of the P element disrupted the function of a vital gene). Of these insertions, 14 failed to complement the *zip[Ebr]* mutant, although in 11 of these cases the phenotype, and presumably the interaction, was weak. Two of these insertions were new alleles of *zipper*, but the third defined a new gene, *RhoGEF2*. Moreover, two previously isolated EMS-induced alleles of the *RhoGEF2* gene also failed to complement the *zip[Ebr]* mutant. Thus the RhoGEF2 gene clearly behaved as an interacting gene. This result was perhaps not surprising given that the role of Rho proteins in remodeling the actin cytoskeleton is firmly established [for a review, see Schmidt and Hall (1998)]. Rho-like proteins have been shown to affect both the actin assembly and the organization of these actin filaments into various actin superstructures.

The finding that mutants in the *RhoGEF2* gene interacted with the *zipper* mutant raised the possibility that mutants in genes encoding other components of the Rho signaling pathway might also interact with *zipper*. Indeed, the two EMS-induced non-complementers (*E3.10* and *J3.8*) recovered in the original deficiency search mapped in the same interval as the RhoA gene. Sequencing of the RhoA gene in flies carrying the *E3.10* and *J3.8* mutants revealed that these mutants were indeed lesions in the RhoA gene.

The screens performed by Halsell and Kiehart (1998) and Halsell et al. (2000) comprise one of the more successful applications of second-site non-complementation. From our point of view, the success of these screens reflected the use of a deficiency collection that allowed the authors to cover much of the genome in a straightforward and facile manner. The coupling of the deficiency screen to a screen for interacting P insertion mutants provided this approach with even greater power.

An example of Type 2 SSNC in the nematode *C. elegans*

Yook et al. (2001) and her collaborators also studied SSNC in genes whose protein products function at the synaptic junctions of the nervous system of *C. elegans*. These proteins regulate the release of neurotransmitters into the synapse by exocytosis, a process required for coordinated movements. They observed that mutations in the genes encoding the physically interacting synaptic proteins UNC-13 and syntaxin/UNC-64 failed to complement one another. In other words, mutants in both genes that were fully recessive produced an uncoordinated phenotype as double heterozygotes. The intriguing component of this study was that a clever drug resistance assay allowed these workers to quantitate the degree of non-complementation rather than depending only on the discrete "uncoordinated versus wildtype" phenotypic assay.

Non-complementation was observed only when at least one mutant encoded a partially functional, but weakly poisonous, gene product. That is to say, although these mutants were recessive in terms of the uncoordinated assay, the more sensitive drug-resistance assay detected a defect even in heterozygotes. This defect was not simply due to a 50% decrease in the amount of wildtype protein, because no defect in the drug resistance assay was observed in heterozygotes for null alleles. Non-complementation did not require partially functional, or poisonous, alleles at both loci. One such allele, when combined with a null allele of the second gene, also produced an uncoordinated phenotype. However, non-complementation was not observed between null alleles of these two genes, and thus this genetic interaction does not occur with a simple decrease in dosage at the two loci. Thus, this genetic interaction requires a poisonous gene product to sensitize the genetic background.

To quote the authors, "hypomorphic mutations . . . which are recessive in behavioral assays, can act as weak poisons as heterozygotes in quantitative drug sensitivity assays. These poisons sensitize the process of neurotransmission to perturbations at other synaptic loci, resulting in nonallelic noncomplementation. In addition, it is the presence of these poisons rather than a simple decrease in the dosage of the gene product that is essential for nonallelic noncomplementation. . . ."

These workers further demonstrated that non-allelic non-complementation was not limited to interacting proteins. Although the strongest effects were observed between loci encoding gene products that bind to one another, interactions were also observed between proteins that do not directly interact but are members of the same complex. The authors explain such long-range effects by noting that "in processes requiring the correct assembly of large protein complexes, a single faulty subunit can render a large number of gene products inactive by participating in and poisoning protein complexes."

Yook et al. (2001) also observed non-complementation between genes that function at distant points in the same pathway, implying that physical interactions are not required for non-allelic non-complementation. Of course, mutations in genes that function in different processes, such as neurotransmitter synthesis or synaptic development, usually do complement one another. Thus, this genetic interaction was specific for genes acting in the same pathway, that is, for genes acting in synaptic vesicle trafficking.

3.4.3   Type 3 SSNC (combined haplo-insufficiency): the interaction is allele-independent at both loci

This type of SSNC neither requires nor implies the physical interaction of the two proteins. Rather, it suggests only that reducing the dosage of the product of gene A is a survivable event for the cell, unless it is further crippled by a reduction in the dosage of gene B. Combined haplo-insufficiency does not require allele-specificity at either gene, and it is sometimes created by using null alleles or deficiencies at one or both genes. This type of SSNC is probably the most common, and sometimes the least interesting. One could imagine, for example, a case where the mutation in gene B simply depressed the rate of transcription of gene A, thus decreasing the level of A protein production below some threshold of function. However, one could also imagine examples where the two proteins do act in functionally related processes. This latter case can be informative in terms of understanding the specific biological process in question.

An example of Type 3 SSNC involving two motor protein genes in flies

The *nod* and *ncd* genes of Drosophila both encode proteins that control the interaction of chromosomes with their microtubule tracks. When homozygous, loss-of-function mutations of both *nod* and *ncd* cause high frequencies of meiotic chromosome misbehavior in females (i.e. homologous chromosomes often fail to segregate from each other). Both *nod* and *ncd* mutants are fully recessive; no effect on meiotic chromosome behavior is seen in *nod*/+ or *ncd*/+ females. However, elevated frequencies of meiotic failure are observed in doubly heterozygous *nod*/+;*ncd*/+ females (Knowles & Hawley 1991).

The types and frequencies of meiotic failures seen in double heterozygotes are more similar to those observed in *nod* homozygotes than they are to those seen in *ncd* homozygotes. On this basis, Knowles and Hawley concluded that heterozygosity for mutants at the *ncd* gene enhances the effects of reducing the dosage of the *nod*+ gene, and thus creates a nod-like phenotype. They went on to show that the same levels of meiotic failure could be observed when either or both of the two mutant alleles was a deficiency. Thus, there is no allele-specificity for either gene. Further studies showed that these two proteins are not co-localized on the meiotic spindle; the Nod protein binds to chromosomes while the Ncd is bound to the spindle. Nor is there any evidence that the proteins physically interact. Rather, it appears that reducing the copy number of the *ncd+* gene creates a spindle that is less tolerant of a reduced dosage of the *nod+* gene.

Other examples of combined haplo-insufficiency in Drosophila can be found in the work of Jackson and Berg (1999) and Kidd et al. (1999). An example of non-allelic non-complementation involving null alleles of the genes encoding the transcription factors MF1 and MFH1 with respect to cardiovascular development in the mouse has been reported by Winnier et al. (1999). A second mouse example involving *hoxb-5* and *hoxb-6* genes has been reported by Rancourt et al. (1995). The *hoxb-5* and *hoxb-6* genes are adjacent in the mouse HoxB locus and are members of the homeotic transcription factor complex that governs establishment of the mammalian body plan. Although loss-of-function mutants at the *hoxb-5* and *hoxb-6* genes are fully recessive, *hoxb-5*, *hoxb-6* transheterozygotes (*hoxb-5– hoxb-6+/hoxb-5+ hoxb-6–*) display a mutant phenotype.

3.4.4   Summary of SSNC

We have listed above three basic types of SSNC. We have also offered a simple set of rules for distinguishing the three types. In Type 1 SSNC, the two mutant proteins physically interact to produce a poisonous protein dimer or complex. This type of genetic interaction is heralded by allele-specificity for both genes, and cannot be observed when examining a null allele at either gene. In Type 2 SSNC, the mutant protein produced by a specific allele at one of the two loci to sequester the reduced level of normal protein produced by the other gene. This type of genetic interaction is heralded by allele-specificity for one of the two genes, and can be observed when examining a null allele at the other gene. In Type 3 SSNC, the phenotype in double heterozygotes results not from any interaction of wildtype or mutant gene products, but rather from combined haplo-insufficiency. This type of genetic interaction is not allele-specific for either gene, and can be observed using null alleles at either or both genes.

Are there cases of second-site non-complementation in humans (digenic inheritance)? At least one human disease, Digenic retinitis pigmentosa, appears to be the result of double heterozygosity for recessive mutations at two unlinked genes: peripherin/RDS and ROM1 (Kajiwara et al. 1994). Mutations in both the peripherin/RDS and ROM1 are fully recessive when present alone, but double heterozygosity has been shown to produce the disease in three separate families. A possible molecular explanation for this interaction is provided by the work of Goldberg and Molday (1996). They showed that the missense peripherin/RDS mutant is conditionally defective with respect to its subunit assembly. Unlike wildtype peripherin/RDS protein, the mutant protein cannot properly assemble into homotetramers on its own. However, the authors were able to show that the mutant peripherin/RDS protein could form a structurally normal heterotetrameric complex in the presence of wildtype ROM1. Presumably then, the cells of the retina can utilize the mutant peripherin/RDS protein as long as sufficient ROM1 is available to facilitate assembly. But loss of one functional copy of ROM1 appears to retard even this process, and thus creates an insufficiency of assembled peripherin/RDS complexes. Investigators have also found a number of human deafness disorders whose inheritance is consistent with second-site non-complementation [cf. Adato et al. (1999)].

Perhaps a similar example of this phenomenon in humans is provided by the unusual genetics of a complex human disorder called Bardet–Biedl syndrome (BBS). Mutations at six genes have been implicated in the causation of this disorder. The genetics of this disorder are exceedingly complex, but considerable data led to the model that this syndrome was inherited as an autosomal recessive disorder and that mutations at any one of six loci might be able to induce the disorder. None the less, the pedigrees were not fully consistent with simple models of autosomal inheritance. There were cases of unaffected individuals who appear to be homozygous for at least one of the six loci, while other similarly homozygous individuals were affected. Recently, Katsanis et al. (2001) have demonstrated that at least some affected individuals are homozygous for a loss-of-function mutation at one locus and heterozygous for a loss-of-function mutation at one of the other five loci. Katsanis et al. (2001) wish to refer to this phenomenon as triallelic inheritance. However, we rather agree with Burghes et al. (2001) in suggesting that this phenomenon is best referred to as "recessive inheritance with a [dominant] modifier of penetrance" ("dominant" was inserted by us). As Burghes et al. (2001) point out, there are other examples of this phenomenon in flies and mammals. In Drosophila, homozygosity for loss-of-function mutation in the fascilin gene is without phenotype unless this genotype is combined with heterozygosity for a mutation in the Abl gene (Elkins et al. 1990).

Pushing the limits: third-site non-complementation? If SSNC works with two genes, why not try it with three? In other words, suppose that heterozygosity for mutants at two genes created a sensitized background, without creating a phenotype. Perhaps then the addition of a third mutant might be enough to create that phenotype. In their studies of the epistatic interactions of genes controlling sex determination in Drosophila, Baker and Ridge (1980) discovered that XX female flies that were simultaneously heterozygous for mutants in the *tra* and *tra-2* genes were wildtype. But the addition of a third sex-determining mutant in the heterozygous conditions created an intermediate sexual phenotype (intersex). Indeed, every known sex-determination mutant was shown to exert a dominant phenotype in the *tra-2/+*, *tra/+* background.

These observations suggested a screen for new mutations in the sex-determination pathway, i.e. by searching for dominant mutants that create an intersex phenotype in the *tra-2/+*, *tra/+* background. Such a screen is described by Belote et al. (1985). The authors note that, "One advantage of such a screen is that it can potentially allow the detection of pleiotropic mutations that affect not only sex determination, but also some vital process and thus are homozygous lethal." In a fashion similar to that described for the Halsell and Kiehart screen above, the authors began by screening a large collection of deficiencies that together encompassed some 30% of the genome. Of 70 deficiencies tested, 12 tested positive in this assay. However, seven of these deficiencies included a *Minute* gene. *Minute* genes encode ribosomal proteins and as heterozygotes have the capacity to enhance many heterozygous mutations (see above). This is simply the consequence of combining a decrease in gene dosage with a further decrease in protein synthetic capability. However, one of the remaining five deficiencies, which removed a small region on the X chromosome, had the strongest effect on their assay. Several independent lines of evidence indicate that this region did indeed contain an important sex-determination gene. We are not aware of subsequent uses of this type of screen. That is probably a pity. It is perhaps unfortunate that a technique with such significant promise has not been more widely used.

3.5   An extension of second-site non-complementation: dominant enhancers

We have noted above that a screen for second-site non-complementation is really a rather specialized form of a screen for dominant enhancers. The only difference is that in a more general screen for dominant enhancers the target mutation need not be heterozygous. One can use a hypomorphic mutation as a homozygote, as a hemizygote on the X chromosome, or even as a transgenic insertion into the genome (with the wildtype copies knocked out) as the target for enhancement. If the choice of the target allele is good and the screen is well defined, this approach can be extremely effective. An example of one such screen is presented below.

3.5.1   A successful screen for dominant enhancers

The technique of screening for dominant enhancers of hypomorphic mutants is now well established in many organisms. Simon et al. (1991) performed an elegant screen for mutations in genes whose products interacted with the protein tyrosine kinase encoded by the X chromosomal *sevenless* (gene) in Drosophila. The product of the *sev* gene specifies a membrane receptor tyrosine kinase whose function is required for the proper differentiation of one type of photoreceptor cell (R7). The R7 cell is present in each of the 800 repeating units (ommatidia) that comprise the fly's compound eye. The *sevenless* gene product appears to be required only for R7 differentiation, and in its absence the cell that would normally become R7 takes on another fate. The result is a fly in which the organization of each of the ommatidia is disrupted, resulting in a compound eye with a rough or disorganized appearance. Thus it is reasonable to propose, as did the authors, that the Sevenless protein "act(s) as a receptor for a signal that determines whether the presumptive R7 cell becomes a photoreceptor."

Simon et al. (1991) noted that the most straightforward way to isolate new mutants in the process of R7 differentiation would be to isolate new recessive mutations in other genes that result in the loss of the R7 cell. Indeed, such screens, done by others, had already isolated two such genes: the *bride of sevenless* gene, which encodes the ligand for Sevenless (Reinke & Zipursky 1988) and *sevenless in absentia* (Carthew & Rubin 1990). None the less, Simon et al. (1991) reckoned that many steps in the sevenless regulatory pathway might not be specific to the R7 cell, but rather might be shared with other signal transduction screens. Mutants in such genes would not specifically affect R7, and indeed they might be lethal as homozygotes. To obviate these difficulties they designed a "more sensitive screen that allows the identification of mutations that *reduce* rather than abolish the activity of downstream signalling proteins" (our italics).

Using a well-studied viral protein tyrosine kinase (v-src) as an example, Simon et al. (1991) used site-directed mutagenesis to create two putative temperature-sensitive mutant alleles of sevenless and transformed these constructs back into Drosophila. In the absence of a wildtype *sevenless* gene, these two mutant constructs did indeed produce a temperature-sensitive loss of the R7 photoreceptor cell, such that when flies carrying one of these mutant alleles were reared at 22.7°C the eyes were wildtype, but when flies of the same genotype were reared at 24.3°C the eyes displayed the mutant phenotype.

The authors now screened for EMS mutations which, as heterozygotes, caused the *sevenless* allele to exhibit a mutant sevenless phenotype at the permissive temperature (22.7°C); i.e. mutations that behaved as dominant enhancers of the temperature-sensitive *sevenless* allele. Think about this, it is really a straightforward F1 screen for "dominant" mutations that produce an easily scored phenotype: rough eyes. From over 30,000 treated chromosomes screened, 20 such *Enhancer of sevenless [E(sev)]* mutations were recovered. These mutants defined seven lethal complementation groups such that transheterozygotes within each group do not survive. The recessive lethality here is important; such mutants would not have been recovered in a direct screen for recessive visible mutations. More critically, none of these mutations produced a rough-eye phenotype in the presence of a wildtype allele of the *sevenless* gene. Thus, for each of the *E(sev)* mutants the ability to produce a phenotype depends on the partially compromised Sevenless protein that was produced by the temperature-sensitive allele.

Two lines of evidence indicate that the genes defined by these mutants do indeed encode shared components of a number of signal transduction pathways. First, by using a clever system to produce mitotic clones of cells homozygous for these mutants in the eye (see chapter 5), the authors bypassed the recessive lethality of these mutations and showed that homozygosity mutations for at least four of these genes appear to reduce the ability of an eye cell to develop as an R7 cell or as any photoreceptor. Thus these genes appear to participate in some process that is common to all of the photoreceptor cells in the developing eye. Second, mutants in these four genes also attenuated the signaling by another protein tyrosine kinase (Ellipse).

Perhaps not surprisingly, these mutants did turn out to define shared components of the signaling process, including the *Ras* gene (which is a component of many signal transduction pathways) and a putative CDC25 homolog. The latter gene had previously been identified both by recessive lethal mutations and by a dominant mutation called *Son of sevenless (Sos)*. The *Sos* mutation had been identified by its ability to suppress a specific allele of *sevenless*. Evidence in other organisms suggested that CDC25 may act to activate the Ras protein, and so the Sos protein may also act by catalysing the activation of Ras. A drawing of the regulatory pathway elucidated from these studies is presented in figure 3.7. The finding of these and other components of the pathway has led to a thorough understanding of the Sevenless signal transduction pathway.

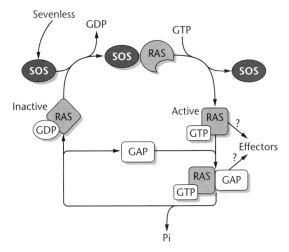

Figure 3.7 Components of the Ras signaling pathway in Drosophila. (Adapted from Simon et al. 1991)

The success of this screen lay in the creation of a "threshold" temperature-sensitive allele that was capable of creating an easily scored phenotype. By excluding enhancers that did not depend on the presence of the sevenless temperature-sensitive mutation to create a phenotype, the authors excluded any unrelated simple dominant mutations. Using mitotic clone analysis to verify a defect of the enhancer mutations alone in R7 cell formation further strengthened the specificity of the screen. The search for dominants allowed the recovery of loss-of-function alleles in essential genes whose products might function in shared pathways.

Summary

The purpose of this chapter was to introduce you to the complementation test and warn you that it can lie to you. Such "lies" are rare and most of the time complementation tests can and should be taken at face value. But if the complementation data are at odds with mapping, or the transheterozygote phenotype is too weak, or the pattern of complementation among multiple alleles is unduly complex, a few "lie-detector" tests to detect intragenic or non-allelic complementation effects are clearly in order.

4 Suppression

We are about to describe a number of cases in which two mutations interact to produce a normal, or near normal, phenotype.[1] These cases are examples of *suppression*.

4.1 A basic definition of genetic suppression

We define suppression as follows:

> Mutant **a1** produces some measurable phenotype in an otherwise wildtype background. However, that phenotype is masked in animals that also carry the **b1** mutation, such that a more nearly normal phenotype is produced. In this instance, the **b1** mutation is said to suppress the phenotypic effects of the **a1** mutant.

Note that we left a fair amount unsaid in this definition of suppression. For instance, we did not tell you whether or not the **b1** mutant creates a phenotype on its own and, if so, whether that phenotype is related to the phenotype exhibited by the **a1** mutant. Nor did we stipulate whether **a1** and **b1** were dominant or recessive, or whether they define the same or different genes. We left these issues unspecified because suppression can encompass virtually all of these possible cases. Both dominant and recessive mutations can be suppressed, and suppressor mutations themselves can be either dominant or recessive. In some cases suppressor mutants are only defined by their ability to suppress some other mutations, while in other cases the suppressor mutants may exert phenotypes of their own. Finally, there are examples in which **a1** and **b1** lie within the same gene (**intragenic suppression**) as well as cases where they lie in separate genes (**extragenic suppression**).

We care about suppression because it can tell us a great deal about the functions of the genes involved, both in terms of the relative functions of the gene products and the mechanisms of gene expression. For example, we will describe below cases in which suppression results from the physical interaction of the two mutant proteins in a fashion that corrects the defect(s) caused by the initial mutation. For this reason, screens for suppressor mutations are a commonly

---

[1]   If your mother ever told you that two wrongs never make a right, you can call her right now and tell her that she was wrong.

used technique to identify genes whose products interact with the product of a given gene of interest.

Unfortunately, not all suppressor mutants define genes whose protein products physically interact with the product of the gene being suppressed. Many suppressor mutations occur in genes whose protein products are involved in various aspects of gene expression. These mutants alter the transcription or translation of the suppressed mutant gene in a way that at least partially restores function. Other suppressor mutants act by altering one or another metabolic or biosynthetic pathways in the cell in a fashion that compensates for the defect caused by the original mutation.

So then you ask: "Can't you just tell me the answer? Like everyone else, I am *only* interested in using suppression as a tool for finding genes whose protein products interact with that of the gene I am currently studying. Must I read these many pages just to be told how to find the suppressors I want?" Much as we discourage such impatience, we can give you an answer. *You will have your best luck when you can identify extragenic suppressors that on their own convey a phenotype similar to that of the mutation you are trying to suppress.* If the rationale for that assertion is patently obvious to you, you can skip to chapter 5. If not, we hope that it will be clear by the end of this chapter.

We begin our discussion of suppression with a consideration of intragenic suppression. We then refocus our attention on extragenic suppression by tRNA modification in *E. coli*, and by the physical interaction of mutant proteins in bacteriophage P22. Finally, we discuss the complex issues of the multiple mechanisms of suppression observed in higher eukaryotes.[2]

4.2 Intragenic suppression (pseudo-reversion)

The term intragenic suppression refers to the case where the two interacting mutations both lie within the same gene. Suppose you have just done a large screen for mutants that suppressed the phenotypic effects of mutant **a1**, and have recovered 1,000 animals that carry **a1** but fail to express the expected phenotype. The first thing you should do is ask whether or not you can easily recover the original **a1** mutant from these animals by one or two crosses. In the case where your suppressor mutant defines a different gene than the **a1** mutant, it should be straightforward to recover the **a1** mutant and its suppressor separately. But if the original **a1** mutation cannot be easily separated from the newly isolated suppressor mutation, then the possibility of intragenic suppression needs to be considered. In this case, we reluctantly (very reluctantly) concede that the fastest determination of this issue can be accomplished by sequencing the mutant and "suppressed" alleles.

---

[2] There are numerous excellent and highly detailed reviews of suppression in both eukaryotic and prokaryotic systems. For example: Hartman and Roth (1973); Hawthorne and Leupold (1974); Guarente (1993); Prelich (1999); and Manson (2000). Our intent here is to be instructive rather than exhaustive, so readers wishing for a broader array of examples are invited to peruse these articles.

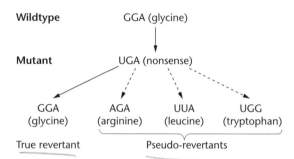

Figure 4.1 Intragenic reversion
of a nonsense mutation

Intragenic suppression of loss-of-function mutations. The most extreme example
of an intragenic suppressor mutation is a true reversion of the original mutation.
As shown in figure 4.1, one could imagine that mutation **a1** is a single base pair
change that creates a nonsense codon midway through the coding region. If
mutation **b1** is a second change that precisely reverses, or reverts, the original
mutant, a wildtype function will be re-established. Suppose instead that mutation **b1** is not a precise reversal of the original **a1** mutation, but rather a change
in another base pair within that same codon allowing the doubly mutant gene
to specify some amino acid (even the wrong one) at that codon. The mRNA produced by this doubly mutant gene can now be translated. If the newly created
codon specifies the same amino acid as the original wildtype codon, a normal
protein will be produced. Indeed, even if the doubly mutant codon specifies a
different amino acid, the resulting protein might still be capable of function
(depending on where the substitution occurs in the protein and the nature of the
amino acid substitution). The critical distinction between these two cases is that
in the latter example, the **a1** and **b1** mutations occur at different sites within the
gene. The **b1** mutant is referred to as a *pseudo-revertant* or a *second-site revertant*.

Second-site intragenic revertants can mediate suppression in several different
fashions. We will consider only three possible cases below. First, we will discuss
the suppression of a frameshift mutation that ablates gene function by disrupting
the translational reading frame downstream of the mutation. Our discussion will
focus on the ability of a second compensatory frameshift mutation to suppress
the effects on the first frameshift mutation by restoring the reading frame. Second,
we will consider cases in which the initial mutation is a missense mutation that
alters the folding of the protein product of this gene. In this case, the ability of
second missense mutation to suppress that mutation results from the ability of
the second amino acid substitution to undo the damage to the protein structure
caused by an original missense mutation. Third, we will consider the suppression
of a dominant poisonous (antimorphic) or neomorphic mutation by a second
mutation that prevents the expression of the poisonous gene product.

4.2.1 Intragenic revertants can mediate translational suppression

*The suppression of frameshift mutations by the addition of a second and compensatory
frameshift mutation.* The first evidence that the genetic code was a triplet came

from studies of intragenic suppressors of mutants within the *rII* gene in phage T4 (Crick et al. 1961). The studies began with a single mutation in the *rII* gene known to have been caused by a mutagen (proflavin) that leads to base insertions or deletions. This mutation was denoted FC0. Putative revertants of this mutation were easily obtained. However, most of these revertants turned out to be due to the occurrence of second mutations in the *rII* gene rather than precise reversals of the original mutation. These additional mutations within the *rII* gene act as suppressors of the original mutations.

As we have noted above, a suppressor mutation can be distinguished from a true revertant because, in the case of a suppressor mutation, it should be possible to re-isolate the original mutation. (*This is especially true for bacteriophage genetics where it is possible to score for recombinants that occur at very low frequencies. It is not practical for most higher eukaryotes.*) Consider the case drawn below, where FC0 is the original mutation and SF is the suppressor mutation. A crossover with a wildtype phage should allow one to re-isolate the original FC0 mutation:

———FC0———SF———
 × → ———FC0—————
—————————————

Using this technique, it was straightforward to demonstrate that these apparent revertants were the result of intragenic suppressor mutations that were themselves frameshift mutants in the rII gene. One can push this trick a bit further, and Crick et al. (1961) did just that. They isolated suppressors of these new frameshift mutants. The suppressors also turned out to be *rII* mutants. Let us refine our nomenclature here. Call the FC0 mutation the archetypal Class I mutant. All suppressors of Class I mutants will be called Class II mutants. All suppressors of Class II mutants will be thought of as Class I mutants, for reasons that will become clear in a moment. Using these terms we can state some rules:

· Every observed case of suppression resulted from combining a Class I mutant and a Class II mutant.
· No pairwise combination of Class I mutants displayed suppression.
· No pairwise combination of Class II mutants displayed suppression.
· But not every combination of Class I and Class II mutants evidenced suppression.

To explain these results, Crick et al. supposed that the gene was read from a defined starting point using a consistent reading frame, such that each set of *n* base pairs encoded one amino acid. If one then adds a single base, the entire frame is thrown off. Consider the model below, in which each underlined grouping of three bases represents the unit in which the code is read, i.e. the codon:

C-A-U-A-A-U-G-A-U-U-G-G-C-G-G-C-A-G-C-A-U-G-G-G-C-A-G-C-A-U-

Now insert a C after the second U:

C-A-U-A-A-U-C-G-A-U-U-G-G-C-G-G-C-A-G-C-A-U-G-G-G-C-A-G-C-A-U-

Assuming that different codons have different meanings, and they must, the whole "sense" of the message will be disrupted after the insertion.

Now imagine a mutation that deleted the third A:

C-A-U-A-U-G-A-U-U-G-G-C-G-G-C-A-G-C-A-U-G-G-G-C-A-G-C-A-U-

Again, from the deletion forward the entire message will be changed.

But now combine the insertion and the deletion:

C-A-U-A-U-C-G-A-U-U-G-G-C-G-G-C-A-G-C-A-U-G-G-G-C-A-G-C-A-U-

Note that the two mutations compensate for each other and restore the reading frame after the second mutation. Those codons including, or separating, the two mutations will be changed, but on either side of the two mutants sense is restored. That the codon was, in fact, read in sets of three bases was demonstrated by the observation that three insertions or three deletions could in some cases restore functionality. (In fact, Crick et al. only really proved that the codon was read in sets of three bases or in multiples of three bases. After all, the insertions and deletions *could* have involved deletions of sets of two or more base pairs. But these mutants have now been sequenced. The original FC0 mutant was indeed a single base insertion, and its suppressors are one base pair deletions. The code is read in sets of three bases.)

It is critical to realize that this experiment made a strong case for the code being degenerate (i.e. that one amino acid can be specified by more than one codon). The fact that one could get rescue at all meant that the out-of-frame sequence between the insertion and deletion had to be readable. This is especially true when one considers that the suppressor mutations were often well separated within the *rII* gene from the original mutant. If 44 of the 64 possible codons were gibberish, suppression would seem highly improbable. Clearly, most possible codons are readable. By definition, since there are only 20 amino acids, there must be more than one codon for some, and in fact, virtually all, amino acids.

So why doesn't combining a single base insertion and a single base deletion always create suppression? Crick et al. (1961) point out that in some cases, "the shift of the reading frame produces some triplets, the reading frame of which is 'unacceptable': for example they may be 'nonsense', or stand for 'end the chain', or be unacceptable in some other way due to the complications in protein structure." Indeed, they were right on both counts. Insertion/deletion combinations that create one of the three stop codons will result in premature chain termination. Similarly, combinations that turn a critical region of the protein into the gibberish that separates the two mutant sites will also fail to display suppression.

4.2.2   *Intragenic suppression as a result of compensatory mutants*

The suppression of missense mutations by the addition of a second and compensatory missense mutation. A fascinating example of this type of "corrective" suppression is provided by the analysis of intragenic suppressors of mutants in the human p53 gene (Brachmann et al. 1998). The p53 protein is a critical regulator of cell division in humans, and the ablation of p53 activity is a characteristic of most human tumors. In most of these tumors the inactivation of p53 is caused by an inherited missense mutant in one of the two p53 genes, and the subsequent somatic loss of the other wildtype copy, a phenomenon known as loss of heterozygosity. These missense mutants can be divided into two classes: mutants in the first class create amino acid substitutions in a critical DNA-binding domain of the p53 protein. These mutations are referred to as "functional mutations." The second class, known as "structural mutations," cause amino acid changes that alter the overall structural integrity and folding of the DNA-binding domain.

Boeke's laboratory proposed that the creation of a second mutation that increased the stability of the folded state of p53 might restore functional activity to at least some of the so-called structural mutants. Using a clever screen for assaying the function of a human p53 gene construct in yeast, Boeke and his collaborators were able to select for second-site intragenic revertants for three tumor-derived p53 missense mutants. In several cases, the effects of these mutations could be understood as the consequence of compensatory structural alteration in the folding of the DNA-binding domain. Other examples of compensatory intragenic suppressors whose action can be understood in terms of known aspects of protein structure can be found in the work of Hong and Spreitzer (1997) on the ribulose-biphosphate carboxylase/oxygenase gene in Chlamydomonas, the work of di Rago et al. (1995) on the yeast mitochondrial cytochrome b gene, and the work of Jung and Spudich (1998) on bacterial rhodopsins. The extension of this technique to the suppression of antimorphic mutants is discussed in box 4.1.

Box 4.1 Intragenic suppression of antimorphic mutations that produce a poisonous protein

Suppose the original mutation that you are trying to suppress is a dominant antimorphic or neomorphic mutation. The easiest way to suppress this mutation is to simply knock out the function of the mutant gene.[3] Suppressing a poisonous mutation is as simple as preventing expression of the mutant gene or rendering the protein product utterly non-functional.

The technique of reverting a dominant mutation was discussed in detail in box 3.4. But as an example, let us once again consider a neomorphic mutant in Drosophila that causes a leg-determining gene to be expressed in the part of an embryo that should become the eye, resulting in an eye → leg transformation. The easiest way to suppress this dominant mutation is by a second mutation that renders the errant leg-determining gene inactive. A small deletion in the coding region, or even a frameshift mutation early in the coding region, would precisely serve this function. There are legions of examples of such phenomena. Indeed, reverting a dominant mutant is often a powerful method for creating true null alleles of a given gene.

---

[3] Think of science fiction movies. No one tries to precisely revert the process that created Godzilla, and turn it back into a small lizard. They just try to kill the mutant lizard before it eats the entire city.

4.3   Extragenic suppression

Perhaps the more interesting case, though, is that of extragenic suppression. How might a mutant in one gene suppress the phenotypic effects of another gene? Unfortunately, there are too many excellent answers to this question. In some cases, suppression can act by "correcting" the mutational defect during (or prior to) translation, thus resulting in the production of a functional protein product.

4.4   Transcriptional suppression

One could imagine various kinds of mutations that affect the ability of a given gene to be transcribed, or the ability of the resulting message to be properly processed and translated. The simplest case might be the effect of an *i*ˢ mutant in *E. coli* which functions as a constitutive repressor of the lactose operon. Such a mutant could easily be suppressed by a deletion of its target DNA sequence, the lac operator. Additional examples in eukaryotes are described below.

4.4.1   Suppression at the level of gene expression

There is a transposable element named *gypsy* in Drosophila that has the capacity to partially or completely inactivate genes by inserting itself between the promoter of that gene and one or more enhancer elements. Many of these insertion mutants are suppressible by mutants in a gene called *su(Hw)* (*suppressor of Hairy wing*). The *su(Hw)* gene encodes a protein (SUHW) that binds to *gypsy* elements and blocks communication between the flanking enhancers and promoters (Dorsett 1993; Kim et al. 1996; Tsai et al. 1997; Scott et al. 1999). Inactivating the *su(Hw)* gene removes this interloper protein and restores communication between the enhancers and promoters (figure 4.2). Thus the inability to produce function Su(Hw) protein allows an *su(Hw)* mutant to suppress the effect of a *gypsy* element on a gene into which it has inserted. [A detailed discussion of the

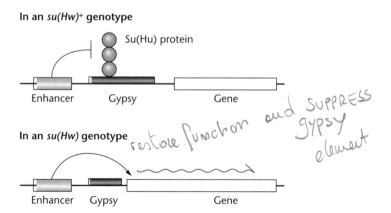

Figure 4.2   The mechanism of suppression by su(Hw) mutants

effects of *gypsy* insertions on gene expression may be found in our considerations of the genetics of the *cut* and *yellow* genes in chapter 6 (section 6.5).] A *gypsy* insertion that actually damaged an enhancer element or inserted into a coding region would not likely be suppressible by *su(Hw)*.

4.4.2   Suppression of transposon insertion mutants by altering the control of mRNA processing

Let us suppose the transposon insertion you wish to suppress is not in some upstream regulatory region, but rather is sitting squarely within the transcribed region of the gene of interest. Now you are out of luck, right? In fact, no. Cases of exactly this type of insertion being suppressible are well known in Drosophila, and the mechanisms by which they are suppressed are instructive.

Some alleles of the Drosophila *vermillion* (*v*) gene are suppressed by recessive mutations at the *suppressor of sable* [*su(s)*] gene. The vermillion gene encodes a protein required for the normal brick-red pigment of the fly eye. In the absence of this protein, the flies have a brighter red phenotype called vermillion. (The *suppressor of sable* gene got its name because mutants in this gene also suppress the phenotypes of another mutation called *sable*.) Previous work has shown that all *v* alleles that are suppressible by the *su(s)* mutations have identical insertions of a large transposon called 412 inserted into the 5′-untranslated region of the *v* gene. Despite the transposon insertion into a non-translated exon, *v* mutant flies do accumulate trace amounts of apparently wildtype-sized *v* gene transcripts in a *su(s)*+ background, and the level of *v* transcript accumulation is increased by *su(s)* mutations (Fridell et al. 1990; Fridell & Searles 1994). This is to say, at some low frequency Drosophila cells are capable of splicing the transposon sequences out of the *v* mRNA, and the frequency of such splicings is increased in cells that are homozygous for the *su(s)* mutation. It turns out that these splicing events are apparently often imprecise. But that is probably tolerable for cells because the insertion occurs in a non-translated exon. Examples of *su(s)* suppression are also known for alleles of other genes. In each of these cases the suppressible mutations are also due to 412 insertions. Thus, a similar mechanism of suppression is likely (Searles & Voelker 1986).

4.4.3   Suppression of nonsense mutants by messenger stabilization in C. elegans

Some years ago Phillip Anderson and Jonathan Hodgkins identified a class of mutants in *C. elegans* that were able to suppress specific alleles of many genes. The suppressible mutations were either nonsense mutations or mutations that resulted in aberrations in the 3′ non-coding region of the transcript. Because these suppressor mutations also caused an alteration in the male genitalia, they were named *smg* (*suppressor with a morphogenetic effect on genitalia*) mutations. To date there are seven genes defined by smg mutations. Surprisingly, these *smg* genes do not encode tRNAs, rather they encode proteins required for a process called mRNA surveillance that leads to the rapid decay of mRNAs containing premature termination codons or altered 3′ non-coding regions (Page et al. 1999). mRNA surveillance, also referred to as "*nonsense-mediated mRNA decay (NMD)*," has been demonstrated in a very large number of organisms. In *C. elegans*, the *smg* mutations block the destruction of messages with premature termination codons. Still, this mechanism raises the following questions: Isn't that nonsense codon still going to block translation? How can the cell tell a premature nonsense codon from a real STOP codon, anyway? The answer to these questions lies in a process known as **translational read-through** (described in more detail below). Ribosomes can read through "out of place" nonsense codons with a low probability, usually by inserting an improperly matched tRNA. As long as the resulting protein can function with this misincorporated amino acid, some degree of function will be restored. The *smg* mutants simply allow mutant messages more time in which read-through can occur.

4.5   Translational suppression

4.5.1   *Simplicity: tRNA suppressors in* E. coli

We begin with a simple case of a nonsense mutant in the *lacZ* gene of *E. coli*. Such a mutant converts a standard "sense" codon that specifies one of 21 amino acids into a STOP codon. The presence of this STOP codon causes the premature termination of the translation of the lacZ message and prevents the formation of its protein product. This protein, β galactosidase, is required for cells to grow if lactose is their only carbon source. Suppose one mutagenizes a population of *E. coli* cells that carry this STOP mutant and then plates these cells on media in which lactose is the only available carbon source. The only cells that will be able to form colonies are those that have regained the ability to make β galactosidase.

Some of these survivors will be true or pseudo-revertants that change the STOP codon into a codon that can specify the incorporation of an amino acid, and thus allow the completion of translation. In rare cases this "reversion event" will create either the original codon or a synonymous codon that specifies the same amino acid as in wildtype. In other cases, the new mutation will create a codon that specifies the insertion of a different amino acid whose presence doesn't impair protein function.

But some of these rare survivors will still carry the original lacZ mutant gene. They survive because they also carry a second "suppressor" mutant in some other gene. Surprisingly, this second suppressor mutant also turns out to be able to suppress nonsense (stop) mutants in several other genes as well. How can a mutation in one gene suppress a nonsense mutant in other genes? More curiously, not all nonsense mutants are suppressible by the same suppressor mutation. Rather, there appear to be three classes of suppressor mutants (known as amber, ochre, and opal suppressors), each of which can suppress a specific set of nonsense mutants.

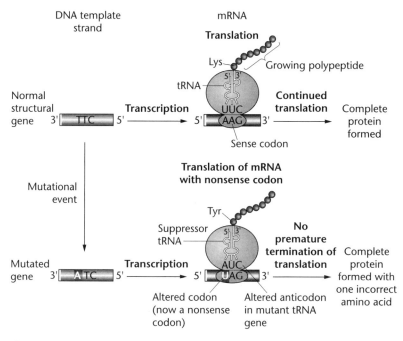

Figure 4.3   A possible mechanism of tRNA suppression in *E. coli*

The answer to this curious puzzle lies in the fact that these suppressor mutations define tRNA genes. The mutation occurs in the portion of the tRNA gene that specifies the anticodon (figure 4.3). This mutation allows the tRNA to recognize one of the three stop codons and insert an amino acid. Even if the inserted acid is not the "correct" amino acid, this replacement will allow translation to continue. If the replacement amino acid does not impair the function of β galactosidase, the substitution will allow the cell to survive on lactose as its sole carbon source. The three classes of suppressor mutations each recognize one of the three types of nonsense codons: UAG (amber), UAA (ochre), and UGA (opal). In each case the suppressor tRNA allows the nonsense codon to direct the insertion of the amino acid carried by that tRNA, and continue translation. Note: this will only work if the directed amino acid substitution is compatible with the protein function.

The demonstration of tRNA-mediated nonsense suppression constitutes one of the more exciting bits of genetic "detective work" of the twentieth century, culminating in a classic paper by Gesteland et al. (1976). An excellent review of our understanding of this process may be found in Murgola (1985). tRNA-mediated nonsense suppression will usually only work when the cell can survive with the small amount of the protein product produced by translation events that utilize the suppressor tRNA. It only works because tRNA genes are functionally redundant and often present in multiple copies. Note that this type of suppression does not involve mutations in genes whose protein products physically interact with or regulate the expression of β galactosidase. Indeed, the suppressor mutations occur in genes (tRNA genes) that don't produce a protein product at all.

But wouldn't a mutation that allows read-through of nonsense codons impair protein production throughout the cell? Actually no, most mRNAs have several STOP codons in each frame at the end of the message. Moreover, it turns out that most suppressor tRNAs don't function very well; they often prevent termination in only a few percent of translation attempts. So while the existence of a suppressor might occasionally tag the replacement amino acid onto some fraction of the translation products, in most cases the changes will be minimal. It turns out that the ability of a given tRNA suppressor mutant to function is dependent on the context in which the mutant stop codon appears (Buckingham 1994).

4.5.2   The numerical and functional redundancy of tRNA genes allowing suppressor mutations to be viable

If a gene encoding a serine-carrying tRNA has now been mutated such that it recognizes a nonsense codon, how do any of the other mRNAs manage to incorporate serine at sites where serine is required? In most organisms, many tRNA genes are present in multiple copies, so there are other tRNA genes specifying the normal serine-incorporating tRNA. Thus removing the ability of one serine-incorporating tRNA gene to direct serine incorporation at a given codon will not greatly impair the cell's ability to insert serine where properly required to do so.

tRNA suppression in *E. coli* provides a lovely paradigm for extragenic suppression, in which a change in a second gene (the tRNA) truly does correct, or substantially ameliorate, the original mutation. Similarly, tRNA suppressor mutants are important tools in yeast (Gesteland et al. 1976) and in *C. elegans* [cf. Wills et al. (1983)]. They have also been used to a lesser extent in more complex organisms such as Arabidopsis (Betzner et al. 1997; Chen et al. 1998), Drosophila (Garza et al. 1990; Washburn & O'Tousa 1992; Robinson & Cooley 1997), and human cell lines (Phillips-Jones et al. 1995; Panchal et al. 1999). Perhaps the most exciting recent example of the use of tRNA suppression in higher organisms is the ability of Leslie Leinwand to demonstrate suppressor tRNA-based suppression in the mouse. Leinwand and her collaborators were able to prevent premature translation of a reporter gene construct carrying an ochre mutation by using a suppressor tRNA in a living mouse (Buvoli et al. 2000). Multicopy tRNA expression plasmids were injected into the heart and skeletal muscles of transgenic mice carrying a reporter gene with an appropriate nonsense mutation. The result was the restoration of translation of a full-length gene product. None the less, tRNA-mediated nonsense suppression has never proven as useful in higher organisms as it has in yeast and prokaryotes, both because of the inefficiency of suppression and because of the deleterious effects exhibited by some potential suppressor mutants.

4.5.3 *Suppression of a frameshift mutation using a mutant tRNA gene*

It is also possible to suppress other kinds of mutants using altered tRNA genes. Indeed, even frameshift mutations are suppressible by the right mutant tRNA. Mendenhall et al. (1987) have identified mutant *S. cerevisiae* glycine tRNA genes that can suppress +1 frameshift mutations in glycine codons. These tRNA mutants were each shown to differ from wildtype at two positions in the anticodon, resulting in the creation of an extended anticodon loop. To expand the utility of this technique, Magliery et al. (2001) have devised a general strategy to select tRNAs with the ability to suppress four base codons from a library of tRNAs with randomized eight or nine nucleotide anticodon loops. The most efficient of these four base codon suppressors had Watson–Crick complementary anticodons.

4.5.4 *Suppressing a nonsense codon using unaltered tRNAs*

As we noted above, ribosomes can read past a nonsense codon at low frequency. For example, Hoja et al. (1998) have shown that a UAG nonsense mutation in the yeast BPL1 gene does not completely block production of the normal length protein. A low read-through rate is accomplished by allowing the ribosome to use the Glutamine (Gln) tRNA, which normally recognizes a CAG codon, to read the UAG nonsense codon. A glutamine is inserted at the site of the nonsense and translation is allowed to continue. Increasing the copy number of this normal Gln tRNA gene in these yeast cells fully suppresses the effects of the original nonsense mutant on cell growth. Samson et al. (1995) and Erdman et al. (1996) have documented other examples of this phenomenon in Drosophila.

   Translational read-through of premature nonsense codons is a bit of an enigma. The frequency with which read-through occurs for any given nonsense mutant is dependent on the sequence, or context, in which the mutant stop codon resides [cf. Kopczynski et al. (1992)]. Read-through is also usually extremely inefficient. Thus, it is not likely to be a problem with respect to proper stop codons, which are often repeated several times in a proper context. None the less, as long as read-through occurs at all, its existence suggests a mechanism by which nonsense mutants might be suppressed. If one could just keep those mutant messages around for a while, and give read-through a chance to make even *some* protein, the cell might be able to survive on a small amount. Exactly such a mechanism of suppression was identified in *C. elegans*, and was described above.

4.6 Suppression by post-translational modification

As noted by Prelich (1999), there are a number of cases in which mutations in a given protein can be suppressed by mutations in a gene whose product is involved in the post-translational modification of that protein. Examples include kinases and phosphatases, which alter the activity of a given protein. A good example is the ability of the *mik1* kinase, when overexpressed, to suppress the effects of a specific point mutant in the *cdc2* gene in *S. pombe* (Lundgren et al. 1991). Obviously, this type of suppression is restricted to the suppression of missense mutations, and null mutations cannot be suppressed in this fashion.

4.7 Extragenic suppression as a result of protein–protein interaction

In one of the truly classic papers in genetic analysis, Jarvik and Botstein (1975) examined the extragenic suppression of three types of conditional mutants in the bacteriophage P22. They began with phage carrying an existing temperature- or cold-sensitive mutation and selected for phage that had regained the ability to grow at the restrictive temperature. For example, starting with a phage carrying a (high) temperature-sensitive mutation, they selected for the production of rare progeny phage that could grow at the high (restrictive temperature). In many cases, these surviving phage were shown to carry a second mutation in another gene. This second mutation not only suppressed the (high) temperature-sensitive defect of the first mutation, but often also conferred cold-sensitivity on the phage (both by itself and in combination with the original temperature-sensitive mutation).

Most critically, these suppressor mutants were almost always in genes whose products are known to physically interact with the product of the gene defined by the first mutation. For example, mutants in gene 5, which encodes a component of the phage capsid, are suppressed by mutants in a gene whose product physically associates with the capsid. To explain this phenomenon, Jarvik and Botstein suggest that the first and second mutations define sites at which the two proteins interact. Quoting these authors, the first "mutation destroys or distorts an alteration, and the suppressor produces a *compensating alteration* that restores the interaction." Because these changes are thought to reflect steric compensations, this type of suppression is often referred to as "*conformational suppression.*" If you think about this carefully, you will realize that conceptually this interaction is a "mirror image" of the type of allele-specific second-site non-complementation between the yeast alpha-tubulin and beta-tubulin genes discussed in the previous chapter. Moreover, the rules for this type of suppression are similar to those for Type I SSNC.

At the end of their paper, Jarvik and Botstein make a compelling argument that the search for allele-specific second-site suppressors should be a useful tool for identifying genes encoding proteins that physically interact with a given protein in higher organisms. They point out that "reversion by suppression is intrinsically function-specific, since the new mutation must remedy a very particular defect caused by the parental mutation." They also carefully noted that, "we do not expect that every suppressor would affect a protein that is intimately related to the parental gene product, but we do suggest that a substantial fraction would." So, if only to test their prognostic abilities, we will now turn our attention to suppression in higher organisms, looking for cases in which the suppressor and the mutant define interacting proteins. We will begin by considering only suppressor screens that identify proteins that physically interact. We will then ask whether or not these interactions reflect true conformation suppression. [The reader in search of a detailed protocol for the identification of such suppressor mutants in yeast is referred to the excellent review by Appling (1999).]

4.7.1   *Searching for suppressors that act by protein–protein interaction in eukaryotes*

A screen for such suppressors in eukaryotes requires rather advanced genetic tools. One needs the availability of a missense mutant to use as "bait" in the suppressor screen. In the best of instances, this mutant should have a conditional or threshold phenotype so that changes in the putative suppressor protein can be accommodated without ablating the function of that protein. One needs the ability to perform very large-scale screens and to easily genetically characterize the recovered suppressors. The minimum requirements are:

1 Both the suppressed and the suppressor mutations should have phenotypes on their own, and in the best case the phenotypes of these two mutations should be similar.

2 Suppression should be allele-specific at both loci and cannot involve a null mutant at either locus. Ideally, one also wants facile techniques for eliminating intragenic suppressor mutants from consideration, and facile secondary screen to identify suppressors whose effect is specific to the mutant in question.

3 The existence of straightforward genetic methods for identifying (and usually eliminating) intragenic suppressors from consideration.

4 The existence of efficient secondary screens to sort out the types of suppressors most deserving of further characterization. For example, dominant suppressors with a recessive phenotype similar to that of the mutant being expressed are good candidates because they are likely to define genes whose products are related functionally. Alternatively, the mutant to be suppressed is often pleiotropic, allowing one to select suppressors based on one phenotype and then screen among those suppressors for ones that also correct some or all of the other phenotypes.

To illustrate the difficulties inherent in such a screen, we will present two extraordinary examples of the use of suppression to identify proteins that physically interact and to study the nature of those interactions.

Actin and fimbrin

The interaction of mutant forms of the actin and fimbrin proteins was considered for a long time to be the best example of conformational suppression in yeast. This story began when Alison Adams and David Botstein screened for dominant allele-specific suppressors of a temperature-sensitive actin (*act1-1*) mutation in yeast (Adams & Botstein 1989). They recovered 29 such suppressors, 24 of which defined a single gene called *SAC6* (*suppressor of actin* mutants).

The recovered *sac6* mutants demonstrated an allele-specific interaction with *act1* mutants because they suppressed only certain *act1* alleles and not others. Three of the 16 *sac6* mutants tested were also temperature-sensitive in an *Act1+* background (i.e. in the presence of a normal actin gene). Although both *act1-1 SAC6+* yeast and *ACT1+ sac6* yeast are temperature-sensitive, the double mutant *act1-1 sac6* grows well at the higher temperature. Thus, combining two mutants, both of which are temperature-sensitive, *restored* the ability to grow at the restrictive temperature.

The finding that *sac6* mutants could suppress *act1* mutants was soon followed by the discovery that null alleles of *sac6* display allele-specific second-site non-complementation with the *act1-4* allele (Welch et al. 1993). Moreover, the *SAC6* protein had already been identified as an actin-binding protein by biochemical studies (Drubin et al. 1988). All lines of evidence suggested the ability of *sac6* mutants to suppress *act1* mutants was a reflection of the physical interaction between actin and fimbrin, and thus possibly might be an example of conformational suppression.

This suggestion of conformational interaction was strengthened by the results of a screen for suppressors of the *sac6* temperature-sensitive mutations that yielded new temperature-sensitive mutations in the *act1* gene. As noted by Prelich (1999), this "mutual suppression" in which mutants in either of the two genes can suppress mutations in the other is a "genetic phenomena frequently associated with interacting proteins." The *SAC6* gene was later shown to encode fimbrin, an actin filament-binding protein. Honts et al. (1994) would go on to show that eight of those *act1* mutants that can be suppressed by *sac6* mutants identify a small region on the actin protein (perhaps the fimbrin-binding site, see below). Moreover, these *act1* mutants show a weakened interaction with SAC6 protein *in vitro*.

Adams and her collaborators (Sandrock et al. 1999) further demonstrated the overexpression of *SAC6* was lethal to the yeast cell and this effect could be specifically suppressed by mutations in the *ACT1* gene. Indeed, the specificity here was truly impressive. Out of 1,326 suppressors they recovered, 1,324 simply reduced the expression of the *SAC6* gene. The remaining two suppressors were both missense mutants in the *act1* gene. Moreover, these two mutants alter amino acids within the same small region of actin that was identified by sequencing the act1 alleles that are suppressible by *sac6* mutants.

Finally, Adams and her collaborators (Sandrock et al. 1997, 1999) would physically demonstrate that those mutant SAC6 proteins that suppressed the mutant ACT1 protein bound more tightly to mutant ACT1 than did wildtype SAC6 protein. Thus, one could think of the actin–fibrin interaction as the metaphorical lock-and-key mechanism in which a mutant key (fibrin) is able to interact only with a mutant lock (actin). However, quite sadly, in at least this case that explanation is likely to be wrong. It turns out that those mutant SAC6 proteins that suppressed the *act1* mutants also bound more tightly to wildtype ACT1 than did wildtype SAC6 protein.[4]

Clearly, the ability of the altered SAC6 protein to bind more tightly to the mutant actin is not specific to the alteration in actin. Were this true lock-and-key suppression, one would not have expected the suppressing SAC6 protein to bind more tightly to wildtype actin than does normal SAC6. Quite the reverse, the suppressing SAC6 protein should bind mutant actin far better than it binds normal actin. Moreover, the sites altered in the suppressing *sac6* mutants are not normally in contact with actin (Hanein et al. 1998).

So, we are left with a simple explanation for all of this that does not really involve a "modified lock opened by a reconfigured key." Simply put: mutants that reduce the ability of actin to bind fimbrin can be suppressed by any one of a number of mutants in fimbrin that increase the affinity of fimbrin for its partner. These mutants do not specifically increase the affinity of fimbrin for the mutant actin, but rather increase the affinity of fimbrin for actin in general. In this sense, these suppressor studies did not identify both the lock and the key. However, Adams and her colleagues most likely identified the fimbrin-binding site on actin (the lock), and that surely was a major accomplishment. We can leave the rest to the crystallographers.

Mediator proteins and RNA polymerase II in yeast

The control of transcription in eukaryotic cells occurs at multiple levels. First, appropriate transcription factors must bind to the gene that is to be regulated. For simplicity, we can think of such transcriptional regulators as consisting of two functional domains. The first domain is a DNA-binding or targeting domain that localizes the transcriptional regulator to the gene of interest. The second domain is an activation domain that either facilitates the formation of a pre-initiation complex, thus allowing transcriptional elongation, or results in the modification of chromatin structure. In the last decade, Rick Young and his collaborators have been able to identify multiple protein co-factors that directly influence the activity of RNA polymerase II (RNAP II) by screening for suppressors of a unique class of internal deletion mutants in the RNAP II.

In *S. cerevisiae* the largest subunit of RNAP II is encoded by the RPB1 gene. The carboxy-terminal domain (CTD) of this protein contains 26 or 27 repeats of a seven amino acid sequence: Tyr–Ser–Pro–Thr–Ser–Pro–Ser. The CTD appears to play critical roles in both the initiation and elongation phases of transcription; reversible phosphorylation of the CTD plays an important role in regulating the activity of RNAP II in transcription. Although the unphosphorylated form appears to be essential for the recruitment of the pre-initiation complex, phosphorylation of the CTD is required for the transition from initiation to

---

[4]   This experiment, which eliminated the "mutant lock meets mutant key" model of suppression, is the ultimate case of a very careful scientist doing just one too many control experiments and killing a really cool hypothesis in the process!

elongation, and for the co-transcriptional processing of the primary transcript [reviewed by Kang and Dahmus (1995).]

Deletions within the RPB1 gene have been recovered that reduce the number of these CTD repeats. These mutants are referred to as *rpb1* mutants (Nonet & Young 1989). Although deletions that retain at least 13 repeat units are fully viable, deletion mutants that retain less than ten repeats are lethal. However, those that retain only 10–12 repeats are both cold-sensitive and unable to grow on inositol as a carbon source. Both the cold sensitivity and the inositol auxotrophy result from the inability of the mutant CTD to interact with the activator portions of various transcription factors, and thus to properly modulate gene expression.

To identify proteins involved in CTD function, Nonet and Young (1989) selected suppressors of the cold-sensitivity of one of three CTD truncation mutants: *rpb1-101* and *rpb1-104*, both of which retain 11 copies of the repeat, and *rpb1-103*, which retains but ten copies of the repeat. We should understand what they are assuming, and what they are asking for, in this selection. The first assumption is that there will be proteins that modulate the interaction between the CTD of RNAP II and the transcriptional activator proteins. The second assumption is that by reducing the repeat number in the CTD to ten or 11 copies they created a threshold where these accessory proteins can bind the altered CTD effectively at normal temperatures, but not in the cold. At this threshold these proteins are presumed to be capable of executing some transcriptional activation events but not others (for example, the induction of those genes required for growth on inositol). The third assumption is that one can create mutants in these accessory proteins that can bind more effectively to the truncated CTD, and in doing so reverse both the cold sensitivity and the inositol auxotrophy.

The first step in this selection was to select for suppressors of the cold-sensitivity of the three *rpb1* mutants. The second step was the weeding out of intragenic revertants. The third step was testing for the restoration of the ability to grow on inositol. One also expected these mutants to be at least semidominant, so they needed to be tested as heterozygotes. (The expectation of semidominance here is important. If the altered accessory protein can effectively mediate the interaction between the transcriptional activator and the mutant RNAP II protein, it ought to be able to do so even in the presence of a normal copy of the protein.) We can also predict that the mutants will display an interesting allele-specificity with respect to their ability to suppress other *rpb1* mutant. These mutant accessory proteins ought to be able to suppress other partial CTD repeat deletions, but should not be able to suppress other types of mutants in the RBP gene. Finally, at least in the best of worlds, one would also predict that these suppressor mutants should have a phenotype similar to that exhibited by the CTD partial repeat deletion mutants.

Young and his collaborators did this screen in rather an interesting fashion. They isolated suppressors in a strain that carried the deleted RNAP II gene on a plasmid and a non-functional large deletion on the chromosome. This trick allowed them to easily re-extract the CTD deletion-bearing plasmid from the strain after mutagenesis, and thus easily distinguish *intragenic* from *extragenic* suppressors. In the case of intragenic suppressors, the CTD deletion-bearing plasmid will no longer be able to produce the cold-sensitive mutant phenotype when transferred back into the unmutagenized parental line (it will now be viable at the restrictive temperature). But for extragenic suppressor mutations, the CTD deletion-bearing plasmid will be unaltered.

Nonet and Young recovered 52 new suppressor mutants, of which 46 were fully characterized. The largest class of suppressor mutants (31 out of 46) were petite mutants. Petite mutations are usually the result of gross deletions in the mitochondrial DNA. These mutants presumably function as suppressors because they slow down the growth of the yeast cells at the restrictive temperature enough to allow the residual function of the mutant CTD domain to be sufficient. But remember, they didn't screen for mutants in genes that interact with RNAP II, they screened for *anything* that would suppress the cold-sensitivity of the CTD deletion mutant.[5]

Unfortunately, 13 of the 15 remaining suppressor mutants were shown to be intragenic suppressors. Perhaps not surprisingly, most (ten out of 13) of these intragenic mutations simply enlarged the repeat domain by duplicating various portions of the repeat coding sequence, and thus increasing the repeat copy number by two to five repeats. Such mutants are more properly considered to be partial revertants than suppressors. The other three intragenic suppressing mutations are independent, but identical, point mutations in a codon within a very highly conserved segment of the RNAP II. All three are G → T transversions at bp 4,953 that change amino acid 1,428 from valine to phenylalanine. It is not clear, but obviously of some interest, why such a mutation could suppress the repeat deletions of both *rpb1-101* and *rpb1-104*.

Discarding these mutants left Nonet and Young with but two remaining candidates for extragenic suppressor mutations. One of these (denoted s3) was too weak a suppressor to study further.[6]

Fortunately, the remaining mutant (s45) was exactly what these workers were looking for. It suppressed both the cold sensitivity and the inositol auxotrophy. Moreover, this mutant was also semidominant. That is to say, it was capable of suppressing the CTD deletion defect in both haploids and as a heterozygote. This mutation was said to define a gene known as SRB2 (*suppressor of RNA polymerase B*), and the mutant was named *srb2-1*. The *srb2-1* mutation was allele-specific in that it showed specificity for the suppression of CTD internal deletions, but did not suppress point mutations in other regions of the RNAP II gene. Moreover, a deletion of the SRB2 gene caused phenotypes similar to those of the original partial CTD deletion mutants. These findings suggested that the SRB2 protein might functionally interact with the CTD of RNAP II protein.

Pushing this successful approach to its limit, Young and colleagues isolated and analyzed additional suppressors [for an excellent review, see Carlson (1997)]. Nearly 30 dominant suppressors were isolated, all of which turned out to be alleles of four genes: *SRB2*, *SRB4*, *SRB5*, and *SRB6* (Thompson et al. 1993). All showed a pattern of suppression similar to that observed for the original *srb2-1* mutation. Biochemical analysis would eventually demonstrate that *SRB2*,

---

[5]   Okay, let's not get discouraged. There are still 15 suppressor mutants left.

[6]   And then there was one. . . .

SRB4, SRB5, and SRB6 proteins do indeed co-purify with a multisubunit complex termed "the mediator" (Thompson et al. 1993). The physical interaction of TATA-binding protein (TBP) with SRB proteins was also reported (Thompson et al. 1993). So in this case the screen clearly worked and was enormously informative.

So why did this screen work so incredibly well? The answer lies in the fact that the selection was well designed from the beginning, and that appropriate secondary screens were able to quickly weed out the less interesting intragenic or petite mutants. But again, this suppression virtually certainly does not reflect an "altered lock–altered key" type of suppression. Rather, all the investigators were seeking were proteins that had a higher affinity for binding the remaining repeat units.

4.7.2   *Extragenic suppression as a result of "lock-and-key" conformational suppression*

> "This rarest of rare circumstances constitutes the signal example of a modified lock opened by a reconfigured key, a mechanism that is prevalent in textbooks, but exceptional in the laboratory."
>
> Manson (2000)

In both of the cases discussed above, the allele-specific suppression reflected protein–protein interactions without requiring lock-and-key conformational suppression. Rather, suppression was the result of compensatory mutations that increased the affinity of the suppressing protein for its mutant partner. All of this raises the question: does true lock-and-key suppression actually occur, or more properly has it been documented, in any organism? Indeed, a small number of examples of this type of conformational suppression have been described in a variety of bacterial systems [for a review, see Manson (2000)]. One of these, which involves the flagellar motor of *E. coli*, deserves some attention.

The flagellar motor of *E. coli* contains two primary parts: the motor element, which is encoded by the *Fli-G* gene, and the stator element, which is encoded by the *MotA* gene. Force generation required the interaction of charged residues on each of these two proteins, specifically arg-90 and Glu-98 in *MotA* and Arg-281, Asp-288, and Asp-289 in Fli-G. In a truly elegant piece of work, Zhou et al. (1998) demonstrated that mutants that reverse one of these charges on the *MotA* can be suppressed by a corresponding and compensatory change on the *Fli-G* protein. In this case, the suppressor protein is incapable of function when opposite a wildtype partner. Thus, the requirements are satisfied for an altered key that only fits an altered lock (figure 4.4).

4.8   Suppression without physical interaction

The examples presented above have dealt with the suppression of recessive mutations in the homo- or hemizygous state. Obviously, the suppression of

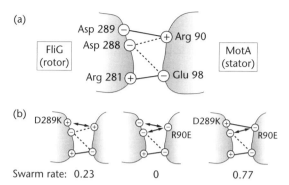

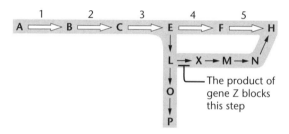

Swarm rate: 0.23   0   0.77

Figure 4.4 Conformational suppression: (a) wildtype; (b) suppression. [Adapted from Figure 8 in Manson (2000)]

Figure 4.5 Bypass suppression. Product H is usually made by steps 1 to 5, beginning with substrate A. A shunt off this pathway produces compound P. There is a branch off that pathway that can also make product H. The use of this branch is usually blocked by the product of gene Z. Mutants in gene Z can thus suppress mutants that block steps 4 or 5.

poisonous anti- or neomorphic mutations provides a powerful tool for the identification of partner proteins or of proteins functioning in the same pathway.

4.8.1   Bypass suppression

A more common type of suppression in higher organisms is a phenomenon referred to as "bypass suppression." In this case, the second mutation simply allows the cell or organism to bypass a defect caused by the first mutation. In its simplest sense, one could imagine that the pathway shown in figure 4.5 is blocked by some loss-of-function mutant in step five, leading to the absence of the final product (H) and the accumulation of an intermediate (E). Imagine a mutant in another gene (Z) that allows the accumulating intermediate E to be shunted into some related but alternative biochemical pathway that produces the necessary product. This type of bypass suppression is well characterized in prokaryotes and played a major role in identifying the various recombination

Box 4.2   Bypass suppression of a telomere defect in the yeast *S. pombe*

There is an elegant and instructive case of bypass suppression in yeast (*S. pombe*) in which a mutant that blocked proper telomere function was suppressed by circularization of all three chromosomes (Baumann & Cech 2001). The mutant *pot1* (*protection of telomeres*) causes rapid loss of telomeres, and resulting chromosome instability. Thus, if one sporulates a diploid that is heterozygous for a *pot1* mutant, the resulting mutant-bearing spores produce small colonies that grow quite slowly. For the next ten generations, these colonies were comprised primarily of very elongated cells that were unable to divide further. Cytological studies revealed that when cell division occurred, one of the two daughters often failed to inherit any DNA.

None the less, after this period of instability the cells in these colonies eventually stabilize and then are able to propagate and divide in the presence of the defect. Indeed, after 75 generations the colony and cell morphology returned to normal. But these cells still lacked functional telomeres, and the *pot1* mutant was very much still there. How then did they suppress this defect? The answer lay in the occurrence of ring chromosome derivates resulting from chromosome breakage events near the ends of chromosomes. Ring chromosomes, by definition, do not have ends and thus do not have telomeres. Once three such events had occurred, all three chromosomes of *S. pombe* were organized as rings. The telomere functions are then all but vestigal. The cell got around a telomere-defective mutant by rearranging chromosomes in such a way as to alleviate the need for telomeres at all. This has to be the defining case of "bypass suppression."

pathways in *E. coli*. One of the nicest examples of bypass suppression in eukaryotes occurs in the yeast *S. pombe*, in which a mutant that blocked proper telomere function was suppressed by circularization of all three chromosomes [Baumann and Cech (2001); see box 4.2].

There is a bit of a tendency among geneticists to view bypass suppression as less "interesting" than conformational suppression. However, it strikes us that the kinds of mutants recovered in bypass suppression, and thus the new genes defined, can be as interesting as those identified by screens for protein–protein interactions. Conformation suppression should only detect proteins that physically interact with the suppressed mutant protein. But bypass suppression can identify a number of genes in related biological pathways. Thus, this technique simply has a wider net and captures related functions. Moreover, to observe conformational suppression, one needs exactly the right pair of mutants. But bypass suppression is not likely to be allele-specific, and *can* involve a null mutant at either locus. Indeed, the ability to suppress a nonsense mutant is a hallmark of bypass suppression. Thus one can identify more genes in the process of a single screen.

4.8.2   "Push me, pull you" bypass selection by counterbalancing of opposite activities

Bypass suppression is often observed when one is screening for suppressors of dominant antimorphic or neomorphic mutants. In this case the second mutant need only block or compensate for the negative effects caused by the first mutant. An example of this effect is provided by studies of the antagonistic activities of two classes of motor proteins in the process of nuclear division in yeast (Saunders et al. 1997). Two kinesin-related motor proteins, Cin8 and Kip1, play an apparently redundant role in the separation of spindle poles and in spindle elongation. A double temperature-sensitive mutant in both genes displays a spindle collapse during mitosis at the restrictive temperature. However, a similar mutation in a gene encoding a third kinesin-related motor, Kar3, partially suppresses the spindle collapse in *cin8 kip1* mutants. In this case, the Cin8 and Kip1 proteins are required to elongate the spindle, while the Kar3 protein is required to contract the spindle. By balancing the loss of the outward lengthening force with a compensatory loss of the inward force (created by the *kar3* mutation), the phenotype is relieved.

Another example of "push me, pull you" suppression is provided by Matthies et al. (1999) regarding the suppression of a dominant effect of an alpha-tubulin mutation on meiotic chromosome segregation in Drosophila. *D. melanogaster* oocytes heterozygous for mutations in the *alpha-tubulin 67C* gene display defects in centromere positioning during the early phases of meiosis I. The centromeres do not migrate towards the poles, and the chromatin fails to stretch during spindle lengthening. These results suggest that the poleward forces acting at the centromeres are compromised in the *alpha-tub67C* mutants. Proper centromere orientation and chromatin elongation can be restored to oocytes bearing the alpha-tubulin mutant by the presence of a loss-of-function mutation in the *nod* gene, which encodes a kinesin-like protein that serves as a "brake" to prevent the poleward movement of the chromosomes. These results suggest that the accurate segregation of achiasmate chromosomes requires the proper balance of forces acting on the chromosomes during prometaphase. Once again, the loss of some functions required to pull chromosomes towards the poles can be compensated for by the loss of one of the functions required to retard that movement. It is all just a balance of forces, of which there are many examples [cf. Willins et al. (1995); Sharp et al. (2000)].

4.8.3 Extra-copy suppression as a form of bypass suppression

We cannot conclude this chapter without at least mentioning the technique of high-copy (or multicopy) suppression in yeast. The idea is that one can suppress a given mutant defect by the presence of a high number of copies of some gene(s) with a related function (Bender & Pringle 1991). In this sense it is really an example of bypass suppression. The mechanics of this technique involve creating a library of genes to be tested, using as vector a yeast plasmid that is maintained at a high copy number within the cell. [There are problems inherent in constructing such multicopy libraries. The nature of these difficulties and a possible solution have been presented by Ramer et al. (1992).] One can then transform mutant cells with this library and search for transformants that can survive the appropriate restrictive condition (i.e. in which the mutant phenotype is rescued). (One then discards those cases where the rescuing plasmid contains the wildtype copy of the mutant gene that one is trying to suppress, and focuses on the other surviving colonies.) A detailed protocol for the analysis of multicopy suppression in yeast can be found in Appling (1999).

An elegant example of true bypass suppression using high-copy suppression is provided by the work of Doug Bishop and his colleagues. The yeast recombination protein DMC1 is required for the completion of meiotic recombination and of meiotic prophase. In a *dmc1* mutant, recombination intermediates (double-strand breaks or DSBs) accumulate and the yeast cells arrest permanently in prophase. Bishop et al. (1999) isolated genes which, when present at high copy numbers, suppressed the meiotic arrest phenotype conferred by *dmc1* mutations. Among the suppressors recovered, two act by altering the recombination process. In high copy numbers, REC114 suppresses formation of recombination intermediates. (Meiotic arrest only occurs in the presence of unresolved recombination intermediates. As long as recombination is either completed or never begun, prophase will proceed normally in yeast.) However, high copy numbers of the RAD54 gene suppress meiotic arrest and promote the repair of recombination intermediates. Several lines of evidence suggest that although the RAD54 is not required for repair of DSB recombination intermediates during meiosis, it is required for efficient repair of DSBs by an alternative pathway that involves sister chromatid exchange operating primarily in vegetative cells [see also Schmuckli-Maurer and Heyer (2000)]. To quote Bishop et al. (1999), "the ability of RAD54 to promote DMC1-independent recombination is proposed to involve suppression of a constraint that normally promotes recombination between homologous chromatids rather than sisters." Thus, the presence of a high copy number of RAD54 genes appears to bypass the *dmc1* defect, as obviated by shunting those unrepaired recombination intermediates into an alternative pathway.

In some cases high-copy suppression analysis *can* identify proteins that physically interact. An elegant example of the power of this type of suppression is provided by the work of Susan Ferro-Novick and her collaborators on the genetics of vesicle transport between the Golgi and endoplasmic reticulum in yeast. Ferro-Novick began with a temperature-sensitive mutation in the *BET3* gene in yeast. At the restrictive temperature these cells are defective in this process of intracellular transport and die (Jiang et al. 1998). A screen for extra-copy suppressors of this mutant yielded a new gene, BET5. This suppression is specific for *bet3* mutants, as BET5 overexpression does not rescue other mutants defective in ER transport. The BET3 and BET5 proteins turn out to be components of the same protein complex. Ferro-Novick's laboratory also reported a similar high-copy suppression analysis that revealed an interaction between the proteins encoded by the *SEC34* and *SEC35* genes (Kim et al. 1999).

Lillie and Brown (1992, 1998) reported another case of high-copy suppression involving the physical interaction of two motor proteins in yeast. These authors demonstrated that a temperature-sensitive mutant in a myosin gene (MYO2) could be suppression by high-copy numbers of a kinesin-related protein encoded by the *SMY1* gene. More recently, they demonstrated that these two proteins physically interact in the cell (Beningo et al. 2000). These authors have also observed that extra copies of *SMY1* can suppress mutations (*sec2* and *sec4*) that define components of the late secretory pathway.

The use of high-copy suppression can also be adapted to higher organisms. The application of this technique to Drosophila by Rorth and her collaborators was described in detail in chapter 1 (Rorth et al. 1998). Finally, there is a variation of this technique in which one screens for suppressors of the effects caused by a given high-copy plasmid. This involves the isolation of suppressor mutations that obviate the defects caused by having the bait gene present in high copy number. Sandrock et al. (1999) demonstrated the overexpression of *SAC6* was lethal to the yeast cell and that this effect could be specifically suppressed by mutations in the *ACT1* gene. An impressive use of this technique in *S. pombe* has also been reported by Cullen et al. (2000). (However, a possible pitfall of this approach has already been discussed in box 3.3.)

4.9   Suppression of dominant mutations

The isolation of suppressors of dominant mutations offers an opportunity to identify proteins involved in downstream regulatory processes. For example, Karim et al. (1996) performed a screen for genes that function downstream of *Ras1* during Drosophila eye development. They began their screen by using a dominant mutant in the *Ras1* gene that causes the eye to become rough in appearance, and then screened for dominant suppressors and enhancers of this rough eye phenotype. From the 850,000 mutagenized flies screened, they recovered 282 dominant suppressors. Mutations in known components of the Ras signaling pathway and 577 dominant enhancers were isolated (such as the Drosophila homologs of Raf, MEK, MAPK, and protein phosphatase 2A), as were mutations in several novel signaling genes. Some of these mutant genes appear to be general signaling factors that function in other Ras1 pathways, while one seems to be more specific for photoreceptor development.

4.10   Designing your own screen for suppressor mutations

Before beginning your own suppressor screen, we suggest that you consider the following issues.

1   *Carefully choose the mutant you wish to suppress.* Perhaps no step in a suppression screen is as critical as choosing the type of mutant you wish to suppress. A screen for conformational suppression is best done with a missense, while a bypass suppression screen is best done with a null mutant, and so on. Everything you end up getting will depend on the bait you use.

2   *Carefully choose your mutagen.* One would be ill-advised to use transposon, X-rays, or a high-copy library to screen for conformational suppressors. If you are attempting to probe for subtle protein interactions, one wishes to change one, and only one, amino acid at a time. The choice is likely to be EMS. But these same tools can be excellent in screens for various types of bypass suppression.

3 *Think about your selection or screen: what question are you asking?* Just what are you asking the cell to do? What phenotype are you trying to suppress? Are there obvious easy ways the cells could ameliorate the effect (such as growing more slowly)? We urge you to choose the easiest selection you can devise, but to have a way to sort among the suppressors for those that remedy the defect that is most specifically related to the process you are studying. You are going to get lots of mutants if you work hard enough. Think carefully about how to sort them out. We are partial to those suppressors that have an interesting phenotype on their own, or which suppress other aspects of the original mutant phenotypes other than the one used in the initial screen.

Summary and a warning

We hope to have made it clear that there are a large number of genetic interactions that can result in one mutation suppressing another. We also hope you realize that screening for suppressors is a powerful tool to dissect a given biological process. Finally, we hope we have told you how to look for whatever kind of suppressor you seek. The value of whatever screen you create will be determined by the kinds of secondary screens. More than anything else, these secondary screens will help you find the few mutations you probably want from among the many varied types of mutations you will probably get.

But we cannot leave you without a strong warning. David Botstein [quoted in Manson (2000)] noted that, "When you push a genetic screen hard enough, you will almost always get what you are selecting for. However, it will usually not be what you want." Manson (2000) reinforces this point by noting that "it is imperative not to underestimate the potentially bizarre and improbable consequences that can transpire when rigorous genetic selection is maintained for an appropriate length of time." Nowhere in the annals of genetics are these statements truer than in the characterization of suppression and suppressors. So if you work hard enough you will get the suppressor lines you seek; but remember, you aren't really asking the cell "what other proteins interact with my favorite protein?" You are asking, "what genetic tricks can let the cell or organism get around a mutant in your favorite protein?" These are very different questions.

So our best advice is to be careful out there. Crazy things abound, and they will bite you if you don't take care.

Determining when and where genes function

5

The chapters so far have focused on isolating mutants that affect a given biological process and sorting those mutants into genes, or complementation groups. We now need to determine whether or not mutants that produce similar phenotypic effects do so by affecting the same biological pathway. We can sometimes answer that question using a tool called **epistasis**. It might also be important to know at what times during development, and in which tissues, these genes interact. These issues can be addressed using **mosaic analysis**. Both of these approaches allow us to position the function of one gene in reference to another, either in terms of a pathway or in time and space.

5.1 Epistasis: ordering gene function in pathways

Epistasis analysis can be used to determine whether or not two or more genes define components of the same biological pathway. In those cases where the genes defined by the two or more mutants actually function in the same pathway, it can show the order in which the gene products function.

A hypothetical example would be the case where two mutants affect the same process in the same fashion. Imagine that two null mutants have been isolated in Drosophila, both of which reduce the fly's life span by 50% when homozygous. As shown in figure 5.1, there are two pathways that control life span (A and B), each of which contributes equally to lengthening life. Disruption of either pathway reduces life span by 50%. Do these two mutants define genes that act in the same pathway, or do they affect different pathways yet yield the same results?

If you make double homozygotes, you discover that their life span remains at 50% of normal (i.e. the phenotypic effect is no worse than that observed in either single homozygote). Based on the model presented in figure 5.1, we can conclude that both of these genes act in the same pathway, which once blocked by the first mutant cannot be impaired any further by the second mutant. If, in contrast, all of the double homozygotes had crawled out of their pupal cases and died immediately, living only a fraction of their normal life span, we would propose

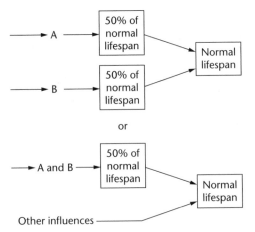

Figure 5.1   Are there one or two pathways that control life span?

that the two genes affect different pathways, both of which contribute independently to life span.[1]

Once it is known that two genes function in the same pathway, we need to determine the order in which they affect it. Rules for ordering genes differ depending upon the type of pathway under analysis. We will begin our discussion of epistasis with the simplest example, the biosynthetic pathway.

5.1.1   *Ordering gene function in a biosynthetic pathway*

Suppose for example we are interested in the biosynthetic pathway described below:

$$A \rightarrow B \rightarrow C \rightarrow D \rightarrow E$$

We have isolated three fully complementing null mutants (*1*, *2*, and *3*) in yeast. All three of these mutants are **E** auxotrophs (they can only grow if **E** is added to their medium). We further discover that mutant *1* can grow if compound **C** is added instead of compound **E**. Mutant *1*'s defect must therefore be prior to, or at, the step that makes the intermediate **C**, because addition of **C** allows the mutant to continue down the pathway and grow.

We now consider the mutant combinations *1,2* and *1,3*. The double mutant *1,2* grows only when compound **C** is supplied. The double mutant *1,3* grows only if compound **E** is added. We can conclude that mutant *2* defines a gene that acts upstream of the synthesis of **C**, while mutant *3* defines a gene that acts downstream of the synthesis of **C**. We cannot yet order genes *1* and *2* with respect to

---

[1]   There is a critical assumption in the preceding example, namely that both mutants were null alleles. Had one or both of these mutants been partial loss-of-function alleles we could not, and should not, have been able to build the model described in figure 5.1; nor could we have interpreted the phenotype of the double homozygotes. If either or both mutants were hypomorphic, then any combined mutant phenotype can reflect the additive sufficiency, or insufficiency, of the gene products, regardless of whether or not the two genes are in the same or different biological processes.

the synthesis of **C** by genetic analysis alone, unless we determine whether or not mutant *1* or mutant *2* can grow in the presence of exogenous compound **B**.

A useful way to consider these data is to think of them in terms of their defects. Redrawing the pathway and placing the genes at the appropriate steps yields:

(*Genes 1 and 2*)   (*Gene 3*)
———————   ———————

$$A \longrightarrow B \longrightarrow C \longrightarrow D \longrightarrow E$$

While, at one level, we might say that the phenotype of each mutant is simply auxotrophic for **E**, in truth we now know more. The information we now have allows us to order the functions of these gene products. The real phenotype of mutants *1,2* is that they cannot make **C**. The phenotype of mutant *3* is that it cannot process **C** to **E**. Thus the real phenotype of *1,3* and *2,3* double mutants is identical to that of mutants *1* and *2* alone; the double mutants cannot make **C**. It doesn't matter how many more downstream mutants (that *could* have made compound **C**) you have, they still cannot make compound **C**. If the cell cannot make **C**, it does not matter if the cell can synthesize **C** → **D** or **D** → **E**, it will not be able to complete the pathway.

Epistasis describes these interactions, and can be defined as the ability of the genotype at one locus to supercede the phenotypic effect of a mutation at another locus. Thus, with respect to the ability to produce compound **C**, mutant *1* and mutant *2* are said to be **epistatic** to mutant *3*. Even though mutant 3 could have made compound **C** by itself, in the presence of either mutant *1* or mutant *2* it can no longer do so.

Experimentally we can define epistasis as follows. For two mutants that produce different but related phenotypes, mutant X is said to be epistatic to mutant Y if the X phenotype predominates in the double mutant organism. A set of **rules for epistasis analysis** of gene order in biosynthetic pathways is listed below.[2]

- *Rule 1*: All mutants involved must be null loss-of-function alleles.
- *Rule 2*: Upstream mutants are epistatic to downstream mutants.
- *Rule 3*: Two null mutants in different genes that fail to show epistasis cannot be in the same pathway.

Excellent examples of the application of epistatic analysis may be found in the study of eye color pigmentation in Drosophila (Phillips & Forrest 1980).

5.1.2   The use of epistasis in non-biosynthetic pathways: determining if two genes act in the same or different pathways

There are many cases, outside of simple biosynthetic pathways, in which one mutant might be epistatic to another. An eye color mutant that blocks any pigment deposition (e.g. white mutants in Drosophila) will be epistatic to mutants

———————

[2] In truth, people do perform epistasis tests with hypomorphs. It often works, but for the reasons considered above, it is risky.

EPISTASIS

EPISTASIS

that affect the color of pigment deposited. Similarly, a wingless mutant will be epistatic to a mutant that produces curled, or otherwise abnormal, wings and an eyeless mutant will be epistatic to a mutant that alters eye color, etc. Although these are rather obvious examples, more sophisticated versions of epistatic analysis are necessary for more complicated biological questions.

Botstein and Maurer (1982) described the canonical example of pathway dissection by discussing mutants that affect the assembly of phage T4. Mutants affecting head assembly have no effect on tail assembly and vice versa (so the processes are presumably distinct), but mutants that block DNA replication are epistatic to mutants in both assembly pathways. Thus one can imagine a pathway that branches after DNA replication into two independent pathways of phage assembly, one for head and one for tail assembly. (This example is also a good warning about over-interpreting the biological mechanisms that underlie these functional dependencies. It is not the case, as might be assumed, that replicated DNA is required for head or tail assembly. Rather, DNA replication is required for the activation of the "late genes" in this phage, whose products are required during head and tail assembly.) All epistasis can tell you is whether or not Step Q requires Step X in order to occur. It cannot tell you the nature of that requirement.

Epistasis analysis can reveal the existence of multiple pathways, even when both (or all) of your mutants define but one of those pathways. An example of this type of analysis comes from the work of Jeff Sekelsky and Kim McKim on meiotic mutants in Drosophila (Sekelsky et al. 1995; McKim et al. 1996). There are two meiotic mutants in Drosophila, *mei-9* and *mei-218*, both of which reduce the frequency of meiotic recombination by more than 90%. These maximum reductions of 90% are observed even with null alleles of these genes. There are really two ways to think about this observation:

1   There is a minor pathway of meiotic recombination in Drosophila that does not require *mei-9* or *mei-218*. Such pathways have been observed in other organisms.

2   There is only one pathway, but there are a pair of structurally or functionally redundant proteins that can substitute for either Mei-9 or Mei-218 proteins with a low probability of success.

The two hypotheses ask a simple question: is the small percentage (10%) of exchange that remains in *mei-9* and *mei-218* mutants the result of a second (independent) pathway that does not require either gene? If this is the case, then we expect the double homozygotes to be no more severe than homozygotes of either mutant alone. The *mei-9 mei-218* double mutant should still show 10% residual recombination. The other possibility is that there is a single pathway in which both the Mei-9 and Mei-218 proteins act separately, with neither mutant being strong enough to fully ablate the function of the pathway. Based on this second model, we would expect the effects of the two mutants on this pathway to be multiplicative, and thus expect only 1% of residual exchange. Sekelsky et al. (1995) observed that the frequency of residual exchange in double homozygotes was not reduced below the level observed in *mei-218* homozygotes alone. Thus, these data support the first hypothesis, in which there are both a major and a minor pathway of meiotic recombination in Drosophila, similar to those observed in other organisms. It is likely that both pathways share a common set of functions, because there are other mutants that completely oblate recombination in Drosophila.

The usefulness of this type of analysis is hard to overstate. As pointed out by Botstein and Maurer (1982), "tests of epistasis allow the grouping of mutations into a pathway structure: two mutations failing to show epistasis cannot be on the same dependent pathway." None the less, our ability to assess epistatic interactions often depends both on the type of mutants we have and on the process or pathway that we are trying to dissect.

if 2 mutations don't show epistasis they cannot be in the same pathway

5.1.3 *The real value of epistasis analysis is in the dissection of regulatory hierarchies*

This section is concerned with the genetic dissection of complex regulatory pathways such as those that determine cell fate or sex. In order to perform this type of analysis, you must have already identified a group of mutants that misdirect the outcome of a given development process (e.g. gain- or loss-of-function mutants that cause genetic males to develop as phenotypic females, or mutants that cause cells that should become bristle-formers to take on some alternative fate, such as becoming neurons). These mutants are rather special; it is not that they simply cause some structure to fail to develop or some process to occur, but rather that they change the direction or outcome of that process.

This analysis also requires that the mutants be fully penetrant, that is to say the phenotype is expressed in every cell or organism, and your collection includes mutants that misdirect the pathway in both directions (i.e. you need mutants to cause genetically male individuals to develop as females *and* mutants that cause genetically female individuals to develop as males). This form of epistasis is somewhat more permissive than the one used for studying biosynthetic pathways, in that we can now use both dominant gain-of-function and recessive loss-of-function alleles (as long as they are fully penetrant). You now need to order the functions of the genes defined by these mutants into a regulatory pathway.

Epistasis analysis using mutants with opposite effects on the phenotype

We begin by considering a hypothetical pathway in the development of a sensory structure in the fly. This structure is built from 16 progenitor cells. The center four cells will become neurons and the remaining (outside) 12 cells are destined to become bristle-forming epithelial cells. During development, the localized presence of an inductive chemical signal (CS) causes the center four cells out of the 16 progenitor cells in this structure to become neurons. In the absence of that CS signal, the remaining 12 cells become epithelial cells. If you prevent the formation of the CS signal, all cells develop as bristle epithelial cells. If you flood the structure with CS, all cells become neurons. Thus the fate of these cells is dependent on a simple binary signal: CS is present *or* CS is absent. We refer to these two situations as **signal states** [following the seminal review of Avery and Wasserman (1992)].

signal STATE (binary sit)
CS yes
CS no

You have isolated dominant constitutive mutants and null loss-of-function mutants that affect this differentiation process in two separate genes (*neu1* and *epi1*). Dominant *Neu1* mutants cause **all** 16 cells to become neurons, even those that do not receive the CS signal. Conversely, null alleles of *neu1* lead to a neuron-less phenotype even in the presence of CS. Dominant *Epi1* mutants also present a neuron-less phenotype, while loss-of-function *epi1* mutants produce the **all** neuron phenotype. To summarize:

Neu1 or *epi1* → ALL neurons
Epi1 or *neu1* → no neurons

It is critical to realize that these two dominant mutants (*Neu1* and *Epi1*) are exerting their effects in two quite different fashions. *Neu1* causes cells that have not received the CS signal to create neurons, while *Epi1* causes cells that have received the CS signal to block neuronal differentiation and become epithelial cells. Thus each of these two mutants affects only one of the two possible signal states (CS present or CS absent). In this case the mutants have opposite effects: *Neu1* turns cells that should become epithelial cells into neurons, while *Epi1* converts cells that should become neurons into epithelial cells.

It is fortunate here that the phenotypes are clean (i.e. fully penetrant) and reciprocal. We can use these mutants to determine whether or not the Neu1 and Epi1 genes act in a single pathway and, if so, determine the order in which they act. The rules for such an analysis are as follows:

1   The pathway we are dissecting must be binary, both in terms of the initiating signal (CS) and the result (neuron versus non-neuron cell fate).
2   Sequential steps in the pathway must act as a series of binary switches.
3   We presume that all mutants tested are either complete loss-of-function alleles or fully penetrant dominants. Each mutant serves to lock a signal switch in the + or − setting.
4   The test is designed to work with two mutants that produce fully opposite effects, i.e. that affect opposite signal states.
5   **When the two mutants affect opposite signal states, the setting of the last "switch" determines the phenotype. The downstream mutant will be epistatic to the upstream mutant.**

We envision a pathway in which the CS signaling leads to the neuronal pattern of differentiation (remember mutational ablation of the CS gene by a *cs−* mutant leads to a neuron-less, all epithelial cell phenotype) by acting through the Neu1 and Epi1 proteins. How would that pathway be constructed? Let us look at the phenotypes of the double mutant combinations:

cs− Neu1 → ALL neurons
Neu1 Epi1 → no neurons

These data tell us that the Neu1 protein acts downstream of the CS signal, and the Epi1 protein acts downstream of the Neu1 protein. Why? We know this because *Neu1* mutants are epistatic to *cs−*. Thus, the Neu1 protein must function downstream of CS in the regulatory cascade. Similarly, because *Epi1* mutants are epistatic to *Neu1*, the Epi1 protein must function downstream of Neu1 in the regulatory cascade.

We can confirm this by observing the phenotypes of two more double mutant combinations:

cs– epi1 → ALL neurons

neu1 epi1 → ALL neurons

[handwritten: EPI1 is critical]

Thus, the Epi1 protein is the critical step in this cascade. We can now understand this process of cell differentiation as consisting of three steps:

1   The function of the Epi1 protein is to prevent neuronal development. We know this because loss-of-function alleles of *epi1* allow all the cells to become neurons and gain-of-function (constitutively activated) alleles of Epi1 cause all cells to become epithelial.

2   The function of the Neu1 protein is to repress the activity (or synthesis) of Epi1. We know this because the loss-of-function *epi1* alleles have the same phenotype as the dominant gain-of-function alleles of Neu1. That is, constitutively expressing Neu1 has the same effect as removing Epi1.

[handwritten: l.o.f. of epi1 same as Neu1 — Neu1 represn epi1]

3   CS activates Neu1: a *cs–* mutant has the same phenotype as a *neu1* mutant, and the overexpression of Neu1 is epistatic to a *cs–* mutant.

In this case we can easily diagram the regulatory hierarchy:

CS → activates Neu1 → represses Epi1 → allows neural development

Thus the most proximate regulatory element is the Epi1 protein which, when left alone, will repress the neuronal choice. Only the interference by a CS-activated Neu1 allows neuronal differentiation. (Note there would have been no point in constructing *Neu1 epi1* or *neu1 Ep1* double homozygotes. In each case both mutants have the identical phenotype, and so presumably will the double mutant.) There are many cases of epistatic interactions leading to the construction of hierarchies. The best studied of these examples are the hierarchies for sex determination in Drosophila and *C. elegans* (Cline & Meyer 1996).[3]

Consider the regulatory hierarchy for sexual development in Drosophila:

Two X chromosomes → Sxl present → Tra present → Dsx female version
One X chromosome → Sxl absent → Tra absent → Dsx male version

In the third step of the sex determination hierarchy, the *dsx* mRNA is spliced to produce a transcription encoding female-producing protein (DsxF) or male-producing protein (DsxM). The direction of that splice is determined by the presence or absence of functional Tra protein. Tra is a protein that contains the splicing of the dsx mRNA molecule. In the presence of Tra, the *dsx* mRNA is spliced to produce DsxF, in the absence of Tra the default splice of the dsx mRNA produces DsxM. The presence or absence of functional Tra is controlled at the level of splicing by the presence or absence of Sxl protein. At the beginning of this hierarchy the presence or absence of Sxl protein is determined by the ratio of X chromosomes to autosomes.

[handwritten: dsx mRNA spliced — DsxF or DsxM]

---

[3]   Perhaps the nicest exposition of epistasis analysis in any higher eukaryote is the study of fly sex determination by Baker and Ridge (1980).

binary
one vs 2X
res F or Male

Sex determination is an ideal case to study epistasis because the initial signal (one X versus two) and the final result (male or female differentiation) are binary. Null and fully penetrant dominant alleles are available for each gene involved in the pathway. The dominant alleles of each gene affect one signal state, while the null alleles of that same gene affect the other. For example, constitutive alleles of *Sxl* will initiate the female pathway of development in the presence of a single X chromosome, while null alleles of *Sxl* will result in XX individuals differentiating as males. The same can be said for constitutive and loss-of-function alleles of *tra*. Dominant mutants of dsx (*dsxM* and *dsxF*) exist that lock the *dsx* gene into producing only DsxM or DsxF protein regardless of the presence or absence of Tra protein. The dominant *dsxM* allele will exert a phenotypic effect only in the XX signal state, while the dominant *dsxF* allele exerts an effect only in the single X signal state. Null alleles of *dsx* produce intersexes.

We can now combine any two mutants that produce opposite sexual transformations, that is to say any two mutants that exert their effects in opposite signal states. For example, the dominant *dsx* mutants will be epistatic to either constitutive or loss-of-function mutants anywhere above them in the hierarchy of XX-induced events. Similarly, null alleles of *dsx* will differentiate as intersex offspring regardless of the genotype with respect to *Sxl* and *Tra*. In every case it will be the phenotype of the downstream mutant (whether loss-of-function or constitutive) that predominates. Remember, it is not possible to use two mutants that have their effects on the same signal state, both mutants will produce the same transformation.

We can also use dominant constitutive mutants in these analyses, as long as their effects are fully penetrant. That is, we could combine a dominant "always ON" allele of Sxl with a loss-of-function *tra* mutant. The result (defined by the phenotype of the downstream mutant *tra*) would be a commitment to male differentiation. All that matters in this pathway is the status of the final binary switch in this pathway, the splicing of the *dsx* gene. If there is no Tra activity, then the dsx gene will be spliced according to the male pathway and male differentiation can proceed.

Epistasis analysis using mutants with the same or similar effects on the final phenotype

EPISTASIS with mutants w/ the same phenotype

The analysis described above requires that the two mutants act on opposite signal states, such that each tested pair of mutants exhibits opposite effects. This is obviously a serious restriction. Many screens are based on recovering mutants that fail to respond to some signal, and hence all mutants have the same phenotypes. In this case, epistatis analysis is possible if one can meet any of three conditions: (1) to obtain opposite-acting conditional mutants of the various genes involved in the pathway; (2) to have available some drug or agent that allows you to block the progression of the pathway under study, without aborting that pathway entirely or killing the cell; or (3) if mutants in the two genes have at least slight differences in terms of their phenotypes. These three conditions are discussed further below.

Table 5.1 Results expected in reciprocal shift experiments using *cs-ts cdc* double mutants. [Adapted from Table 1 in Botstein and Maurer (1982)]

Dependency relationship	Result of shift	
	$17° \rightarrow 37°$	$37° \rightarrow 17°$
Dependent *ts* → *cs* →	+[a]	−[a]
Dependent *cs* → *ts* →	−	+
Independent *ts* →	+	+
cs →		
Interdependent (*cs, ts*) →	−	−

[a]A "+" symbol indicates passage to a second cell cycle (i.e. two arrested cells are found) and a "−" symbol indicates arrest in the first cell cycle (i.e. one arrested cell is found).

The use of opposite-acting conditional mutants to order gene function by reciprocal shift experiments. The reciprocal shift method was first developed by Jarvik and Botstein (1975) for use in the study of phage morphogenesis and then applied to study the yeast cell cycle by Hereford and Hartwell (1974). To do reciprocal shift experiments, you need conditional mutants at both genes that allow you to block the process in two different ways [i.e. such as both *cold-sensitive* (*cs*) and *heat sensitive* (*ts*) mutants for each gene]. As shown in table 5.1, we can learn a great deal by combining a *ts* allele of gene 1 with a *cs* allele of gene 2 (or vice versa).

Assuming a method exists to initially synchronize our population of cells or organisms, we can then allow the cell or organism to go through a process of reciprocal temperature shifts: first a high-temperature phase (inactivating gene 1), then a low-temperature phase (inactivating gene 2). If gene 1 acts before gene 2, this set of temperature shifts will produce a mutant phenotype. Even though gene 2 was functioning during the high-temperature period, the preceding (and presumably essential) prior function of gene 1 was inactivated. The restoration of gene 1 activity during the subsequent shift to cold was of no importance because gene 2 was no longer able to provide the essential subsequent function.

Suppose we reverse the order of temperature shifts (figure 5.2). The initial cold treatment knocks out gene 2, but still allows gene 1 to produce its product. The subsequent shift to high temperature stops gene 1 from functioning but frees up gene 2 to act on the substrate provided by gene 1 during the previous shift. This reverse order of temperature shift is expected to produce a normal, or near wildtype, outcome. This method yields information about whether or not two genes act in the same or different pathways, and about their order if they do function in the same pathway. The power of the method is documented by a schematic diagram (figure 5.3) of the yeast cell cycle pathway as deduced from reciprocal shift experiments (Hartwell et al. 1973; Moir & Botstein 1982).

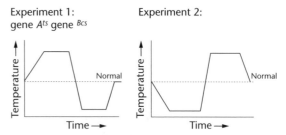

Consider a pathway involving both the
product of gene A and gene B

You have *ts* and *cs* alleles of both genes

Experiment 1:   Experiment 2:
gene *A*^{ts} gene *B*^{cs}

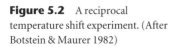

If gene A's function precedes that of gene B
then experiment 2 will restore function

If gene B's function precedes that of gene A
then experiment 1 will restore function

Figure 5.2   A reciprocal
temperature shift experiment. (After
Botstein & Maurer 1982)

Using a drug or agent that stops the pathway at a given point. Suppose you have
only *ts* mutants in genes that affect the completion of the yeast cell cycle.
Obviously, without some set of internal or differentiating landmarks for cell
cycle progression, one cannot either order these genes or assign them to
pathways. Now imagine that you had an agent that blocked the pathway mid-cycle
without killing the cells. You could then test each mutant by comparing the effects
of a regime of high-temperature/drug arrest and release/permissive temperature
with a cycle of permissive temperature/drug arrest and release. High-temperature
ts mutants in genes that act prior to the drug arrest will block cell cycle completion in the high-temperature/drug arrest and release/permissive temperature
cycle, but allow completion in the permissive temperature/drug arrest and

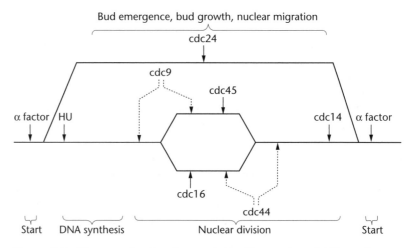

Figure 5.3   The yeast cell cycle pathway as deduced from reciprocal shift experiments

release/high temperature cycle. Mutants in genes that act after the arrest point will exhibit a reciprocal effect. Note: this method only assigns a temporal order, it does not allow you to determine whether or not two genes act in the same pathway.

Exploiting subtle phenotypic differences exhibited by mutants that affect the same signal state to order gene function. Avery and Wasserman (1992) reviewed a case, the cell death cascade in *C. elegans*, in which loss-of-function mutants in two genes (*ced-3* and *ced-1*) exert different effects on the process of programmed cell death and engulfment of the corpse by neighboring cells. Both of these mutants exert their effects in the same signal state, after the cell death pathway has already been switched on.

We can only perform an epistasis analysis in this case because the two mutants have different phenotypic effects in the same signal state. In *ced-3* mutants cell death is completely suppressed. In *ced-1* mutants the cell dies, but the resulting corpse is not engulfed by neighboring cells. In a *ced-1 ced-3* double mutant, the phenotype is identical to that of *ced-3*, and cell death is completely prevented. In this process the outcome is not a single binary state, but rather the pathway branches. That difference in the phenotypes of these mutants, or branching of the pathway in terms of engulfment, allows the analysis.

We can understand this by diagramming the regulatory cascade:

Cell death signal → activates Ced3 $\quad\to$ triggers cell death
$\qquad\qquad\qquad\qquad\qquad\qquad\quad\to$ activates Ced1 → triggers engulfment

In cases such as this, where both your mutants have their effects on the same signal state, the rules of epistasis analysis are different. In this case, **when the two null mutants tested affect the same signal state, the setting of the first switch determines the phenotype.** Thus the *ced-3 ced-1* mutant will be phenotypically identical to *ced-3*.

This analysis becomes more complex if we allow ourselves to use constitutive dominant mutations. A constitutively activated Ced-3 mutant would presumably trigger cell death and engulfment in the absence of signal. However, a cell combining a constitutively activated *Ced-3 ced 1* mutant will still cause death, but would not allow corpse engulfment. In this sense, the downstream mutant is epistatic to the upstream mutant in terms of the engulfment phenotype. However, the Ced-3 mutant is epistatic in terms of the cell death phenotype. It is amusing to imagine the phenotype of a *ced-3 Ced-1* double mutant (where *Ced-1* is a constitutive dominant allele of *ced-1*): living cells trying to induce their neighbors to eat them! Thus, with respect to engulfment, combining dominant constitutive and loss-of-function alleles, it is the downstream mutant that will be epistatic to the upstream mutant. In terms of cell death, it is the upstream phenotype (no death) that is epistatic to the downstream mutant. The issue here is that the rules have changed slightly, because the pathway no longer has a simple binary outcome. A simple regulatory hierarchy now has a complex side branch. How we assess the result depends on which branch we are examining.

5.1.4   *How might an epistasis experiment mislead you?*

Can you be fooled by epistasis? The answer is yes, easily! Avery and Wasserman (1992) describe cases in which the functional orders of gene action were incorrectly obtained because the investigators used partial loss-of-function alleles. (Tools for verifying that your alleles are indeed null mutants were provided in chapter 1.) As noted by Botstein and Maurer (1982), epistasis "tests required detailed knowledge of the mutant phenotype." We would add to that only, "and of the nature of the mutants themselves." Epistasis analysis can also fail if a given component of a regulatory hierarchy has more than two states (ON/OFF), and instead functions more like a dimmer switch. Further, the analysis can fail if you simply misunderstand the type of pathway you are dissecting (Ferguson et al. 1987). If, for example, the final phenotype involves the interaction of parallel or partially redundant cellular pathways, or if unknown back-up or salvage pathways exist, one can easily misunderstand the result. Finally, it is not at all uncommon for the double mutant to have a phenotype quite different from that of either of the two single mutants. Such pitfalls are dangerous and perhaps not all that uncommon. We echo Avery and Wasserman in pointing out that "epistasis analysis alone will not tell you in molecular terms how genes are regulated." All epistasis analysis can do at best is suggest who is regulating whom. We suggest that epistasis is best thought of as a tool that can implicate pairs of genes in the same process. Beyond that, one needs to rely on the tools of molecular or cell biology to elucidate the exact function of the relevant gene products in that process.

5.2   Mosaic analysis: where does a given gene act?

If a given mutation results in a phenotype such as an absent antennal structure, it might seem straightforward to assume that the wildtype product of that gene acts in the tissue primordium that forms that antenna structure. It might be straightforward to assume that, it might also be completely wrong! The gene defined by your mutant may act at a distance by producing a diffusible factor that induces competent cells to take on an antenna fate, or its product might act much earlier in development by controlling the specification of some region or compartment that will eventually be required to make an antenna. How then do we determine when or where a given gene exerts its function? To address the questions of "where," we need to be able to produce and analyze genetic mosaics. A **mosaic** is an animal or plant in which cells or tissues within that single organism have different genotypes and can be distinguished as having a separate origin. The basic question here is very simple: if an organism possesses both wildtype and mutant cells, which cells must be wildtype to produce a normal phenotype? Being able to address that question requires us to accomplish three objectives: (1) we must be able to make the appropriate mosaics at the appropriate time during development (and they must survive long enough to assess the phenotype of interest); (2) we must be able to easily distinguish mutant cells from wildtype cells; and (3) we must be able to assess the phenotype in the whole organisms we produce.

Multiple excellent reviews of mosaic analysis have been written, beginning with Stern (1936, 1968), Nesbitt and Gartler (1971), Hall et al. (1976), Ashburner (1989), Golic (1993), and Greenspan (1997). We lack the space here to cover in detail the rather large variety of methods available for creating mosaics. Rather, we will briefly summarize the techniques themselves, focusing primarily on the types of questions that can be answered by using mosaic analysis. We begin with perhaps the oldest form of creating genetic mosaics: tissue transplantation.

5.2.1   Tissue transplantation studies

In the early 1900s, Ephrussi and Beadle were working on the relationship between genotype and phenotype with respect to eye color in Drosophila. The basic question was: did the genes that produced the color of the fly's eye act in the eye cells themselves? The experiment involves transplanting the larval precursors of the eyes (known as eye imaginal discs, or discs) from one larva to another. When a disc from larva **A** was implanted into larva **B**, the transplanted disc did differentiate into an eye and could be "scored" for eye color in the next stage of fly development in the pupa. One could then ask whether or not transplanting an eye disc from a mutant larva into a wildtype larva could rescue the eye pigment defect. If the surrounding wildtype tissue can rescue the pigmentation of the mutant eye disc, then the necessary gene product for proper pigmentation can be (and probably is) made in cells outside the eye. The issue we are getting here is called **cell autonomy**. *We say that the phenotype of genetic function is cell autonomous if that phenotype is determined by the genotype of the cell or cells that exhibit it.*

So in modern parlance, Ephrussi and Beadle (1936) were asking whether or not eye color was cell autonomous in Drosophila. They were working with two loss-of-function eye color mutants, *vermillion* (*v*) and *cinnabar* (*cn*), both of which produce a fly with bright red eyes. (Normal flies have dark or brick-red eyes as a result of producing both a red and a brown pigment. The wildtype *cinnabar* and wildtype *vermillion* genes are required to produce the brown pigment. Hence, the eyes of flies homo- or hemizygous for either mutant are bright red.) If one transplants an eye disc from a larva that is homozygous for either a *vermillion* or a *cinnabar* mutation into a wildtype larva, the resulting supernumerary eye will be normal in eye color! The wildtype host larva could provide whatever function the *vermillion* or *cinnabar* eye discs were missing. (Think for a moment about the obvious control for this experiment, transplanting *vermillion* eye discs into vermillion larvae, or transplanting *cinnabar* eye discs into *cinnabar* larvae. What is the expected result?) Clearly the functions of the *vermillion* and *cinnabar* genes are not cell or tissue autonomous. These genes must produce diffusible products that played some role in the synthesis of the normal pigment. When the mutant discs were placed into the wildtype host larva, the developing discs were able to obtain these factors from the host tissue and bypass the mutant defect.

Ephrussi and Beadle also used this surgical form of mosaic analysis to ask a second question, namely could one use mosaics to order two genes that function in a given biosynthetic pathway? They noted that the transplantation of *vermillion*

[handwritten margin note:] cell autonomy: a phenotype is cell autonomous if that phenotype is determined by the genotype of the cell

[handwritten margin note:] There is a factor outside the cell involved in eyes color

V into cin = normal

cin. into ver = cinbar
color

discs into *cinnabar* larvae resulted in flies with normal color supernumerary eyes. Thus the *cinnabar* host could provide some component necessary to produce a normal pigment. When eye discs from *cinnabar* larvae were transplanted into *vermillion* larvae, the supernumerary eye remained cinnabar in color. To explain this result Ephrussi and Beadle imagined the brown pigment was the product of a biosynthetic pathway with at least two steps:

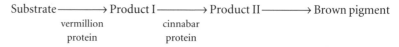

Substrate ———→ Product I ———→ Product II ———→ Brown pigment

    vermillion   cinnabar

      protein   protein

Ephrussi and Beadle argued that the product of the wildtype vermillion gene was the enzyme required to complete the first step, producing Product I, while the product of the wildtype cinnabar gene was the enzyme required to complete the second step, producing Product II. The two mutant defects can be described as a two-step process in which a *vermillion* mutant blocks the first step while a *cinnabar* mutant blocks the second step. One can see how the presence of diffusible Product II in the host larva could rescue either defect. Providing a *vermillion* eye disc with either Product I or Product II will allow wildtype pigment synthesis. Only the rescue of Product II can then rescue the *cinnabar* eye disc. It should also be obvious why a *cinnabar* host can rescue a *vermillion* disc. The *cinnabar* host can make the missing Product I. Make sure you see why a *vermillion* host cannot rescue a *cinnabar* disc. The substrate is tryptophan; Product I is formylkynurenin; and Product II is hydroxykynurenin. The product of the *vermillion* gene is an enzyme named tryptophan pyrrolase that catalyses the conversion of tryptophan to formylkynurenin, and the product of the *cinnabar* gene is an enzyme called kynurenine 3-monooxygenase that catalyses the reaction of formylkynurenin to hydroxykynurenin.

The methods used by Ephrussi and Beadle for making mosaics in Drosophila larvae are not so different from the methods of blastula fusions and blastocyst injections that are commonly used to make genetic mosaics (chimeras) in the mouse or pole cell mosaics in Drosophila. None the less, such transplant mosaics are difficult to create, and thus are often limiting in terms of the types of mosaics one can make. For example, one could not examine the activity of genes in different parts of the eye by simply transplanting in fragments of an eye disc. For this reason two methods of creating genetic mosaics have been developed in Drosophila: high-frequency chromosome loss and mitotic recombination. We begin with a discussion of the use of mitotic chromosome loss.

5.2.2 Loss of the unstable ring X chromosome

The occurrence of occasional flies that are part male and part female (so-called **gynandromorphs**) was observed quite early during the development of Drosophila genetics in the twentieth century (figure 5.4). Thomas Hunt Morgan recovered the first gynandromorph in 1914. A rather large number of papers by various workers demonstrated that these male/female tissue mosaics were indeed genetic mosaics that resulted from the mitotic loss of a single X chromosome during the development of an XX fly (Hall et al. 1976). The resulting XO cells

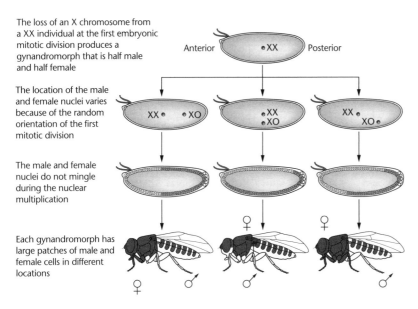

The loss of an X chromosome from a XX individual at the first embryonic mitotic division produces a gynandromorph that is half male and half female

The location of the male and female nuclei varies because of the random orientation of the first mitotic division

The male and female nuclei do not mingle during the nuclear multiplication

Each gynandromorph has large patches of male and female cells in different locations

Figure 5.4   A picture of a gynandromorph

produced clones of male tissue, while the cells that still had two X chromosomes developed as female tissue. (Remember, only the number of X chromosomes carried by each cell, not the presence or absence of a Y chromosome, determines sex in Drosophila. Sex in flies is cell autonomous, a point made by gynandromorphs. The sex of each individual cell is determined solely by its genotype in flies, not by the concentrations of circulating hormones, as in mammals.)

One can imagine how useful such mosaics could be. Suppose, for example, that one could obtain such mosaicism in an initially XX individual that was heterozygous for some X-linked mutant. If that loss event involved the X chromosome carrying the wildtype allele, then the XO male tissue would express only the mutant gene. Provided that one could identify whether or not the cells or tissue of interest were male or female, and which X chromosome had been lost in a gynandromorph, one could ask about the effect of the genotype of those cells on the phenotype of interest. Unfortunately, such spontaneous gynandromorphs are really quite rare. (The dogma is that only one or two in 10,000 XX zygotes will develop as obvious gynandromorphs.) However, the use of mitotic chromosome loss was greatly facilitated by the discovery of unstable X chromosome derivates that underwent mitotic chromosome loss at exceedingly high frequencies. The most useful of these chromosomes is an unstable ring X chromosome, known at $R(1)w^{vC}$ (Hinton 1955, 1957). This ring X chromosome is lost at very high frequency during early fly development, and produces very high frequencies of gynandromorphs. In some cases 20% of the female progeny are recognizable gynandromorphs. While such loss events are observable with appropriate cytological tools (Zalokar et al. 1980; Simpson & Wieschaus 1990), the molecular events that lead to loss remain unclear [for a discussion of possible models, see Hall et al. (1976)].

The lack of understanding of the mechanism(s) of ring loss has not impeded its use as a genetic tool. Mosaics generated by ring-X loss have been used to study a variety of aspects of Drosophila development, especially cell-to-cell communication and behavior. Histological tools were developed that allowed one to distinguish between XO and XX cells in virtually any tissue, and methods were created that allowed the investigator to translocate wildtype copies of autosomal cells of interest to the unstable ring X. (This latter modification made autosomal genes accessible to this type of mosaic analysis.) As these tools developed, more and more cases were accessible to this type of mosaic analysis. The canonical examples of using this system to study cell–cell interactions are the studies of Zipursky and his collaborators (Reinke & Zipursky 1988) on the functions of the *sevenless* and *bride of sevenless* genes in the developing Drosophila eye.

The interested reader is also referred to the classic paper of Hoppe and Greenspan (1986) on the use of gynandromorphs to study cell–cell interactions in the early embryo. These authors studied cell-to-cell interactions involving the neurogenic gene Notch, and provided clear evidence of which cell needed to be wildtype or mutant in order for the embryo to be wildtype. This technique has also been used to study behavior by asking which tissues were required to control specific aspects of various behaviors, including courtship [cf. Hall et al. (1976); Hall (1979)]. Finally, the frequency with which two nuclei in an embryo differ in their genotype after a ring-loss event, can be considered as some metric of the "distance" between those nuclei in the embryo. By computing such "distances" between multiple nuclei, one can create maps of the nuclei in that embryo. One can, for example, position the nuclei that will eventually form eye disc precursors with respect to those nuclei that will eventually become part of cells in the head or thorax. Thorough reviews of this method, known as **fate-mapping**, are provided by Garcia-Bellido and Merriam (1969) and by Hall et al. (1976).

Other mechanisms for inducing mitotic chromosome loss have been identified in Drosophila. Primarily, these methods involve the use of chromosome-destabilizing mutants such as *mit* (Gelbart 1974), *pal* (Baker 1975), *nod* (Zhang & Hawley 1990), and *ncd* (Nelson & Szauter 1992). The utility of these mutants is that they can cause mitotic loss of the Y and 4th chromosomes as well. However, none of these mutants produces mosaics at frequencies as high as does ring-X loss. (*The fact that X and 4th chromosomal loss in Drosophila is tolerated sufficiently to allow mosaic analysis reflects the fact that all cells carrying one, two, or three of these chromosomes are viable.*) None the less, the mutants are still perhaps underused in the sense that they allow the investigator to induce loss of structurally normal chromosomes. The mutants also allow the investigator to control which homolog is lost (the maternal or the paternal).

Similar methods for mosaic analysis have been developed for *C. elegans*. These methods involve the loss of small chromosome fragments (Herman 1984) or of extra-chromosomal arrays (Miller et al. 1996). Several groups of investigators have exploited these tools to study various aspects of cell-to-cell interaction (Kimble & Austin 1989; Seydoux & Greenwald 1989) and cell autonomy in this organism [cf. Villeneuve and Meyer (1990a,b)]. These approaches have an

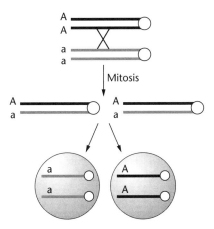

Figure 5.5 Mitotic chromosome recombination between a heterozygous mutant and its centromere *can* produce daughter cells that are each homozygous for one of the two alleles present in the parent cell

advantage over the ring-loss studies in flies, because they do not require the loss of entire chromosomes and thus do not create aneuploidy.

A truly useful tool for creating mosaics in organisms such as flies, zebrafish, and mice would need to avoid this burden of creating aneuploidy. Indeed, the ideal method would not generate aneuploidy at all. That method, genetically controlled mitotic recombination, is described below.

5.2.3   *Mitotic recombination*

As shown in figure 5.5, mitotic chromosome recombination between a heterozygous mutant and its centromere *can* produce daughter cells that are each homozygous for one of the two alleles present in the parent cell. We put the word "can" in italics because only half of the two possible alignments of the two homologs at metaphase will result in homozygous daughter cells. (Remember this is mitosis, the homologs are not going to segregate from each other, even though they recombined.) Since its discovery by Stern (1936), mitotic recombination has long been the favorite tool of fly geneticists both for lineage analysis [cf. Bryant and Schneiderman (1969); Weischaus and Gehring (1976)] and for studies that permit the use of small clones (the later during development the exchange event is induced, the smaller the number of divisions that daughter cells will have remaining, and thus the smaller the clone). There are also methods for increasing the proliferative capacity of a clone relative to its neighbors – thus producing bigger clones – and for genetically marking clones so that they can be distinguished from the majority of parental cells. These techniques are reviewed in Ashburner (1989). Although the frequency of spontaneous mitotic recombination is low, it can be increased by treatment of larvae with X-rays.

Initially, there were real difficulties in the use of mitotic recombination for mosaic analysis. For example, in many cases the mutant one used to identify clones was separate from the mutant one wanted to test in that clone. Since it was not possible to control where exchange occurred, there was a real chance that the clones wouldn't be the ones desired. Moreover, one could not control either the timing or the tissue in which the mitotic exchange events occurred. Timing matters, because a constant worry in studies of induced mosaicism is the **perdurance** of gene products. For example, suppose that a protein product made before the induced exchange was able to persist (or perdure) in the daughter cells of a cell that produced a homozygous mutant clone long enough to create a wildtype outcome. The clone might then appear phenotypically normal, leading the investigator to conclude, quite incorrectly, that a wildtype gene product was not required in those cells to obtain a normal phenotype.

The most useful tool would be a system in which we can control both the frequency and position of mitotic exchange, while also controlling its timing and position during development. Such a system is described in the next section.

5.2.4   *Genetically controllable mitotic recombination: the FLP–FRT system*

The value of mitotic recombination in Drosophila was enhanced enormously by the introduction of the FLP–FRT system by Kent Golic (Golic 1991, 1993; Golic & Golic 1996). This system involves the introduction into Drosophila of a two-component site-specific recombination system (derived from the yeast 2-micron plasmid). In this system, an enzyme (FLP) is uniquely and solely capable of inducing high frequencies of mitotic recombination at a specific site (FRT) (figure 5.6). One can thus use transformation to integrate FRT elements into the fly genome (at one of any number of sites) and an FLP gene whose expression you can control. For example, if your FLP construct was controlled by a heat-shock promoter, you could induce a high frequency of site-specific exchanges at any time in development that you chose, depending only on when you exposed the developing flies to heat shock. The advantages of this system are enormous:

1   The exchanges you induce occur only at the sites where an FRT element is homozygous. Thus, all markers distal to the FRT will co-segregate at the ensuing mitotic division. You can use any visible marker distal to the FRT site to mark your clones.

2   You do not need to use radiation to obtain high levels of exchange.

3   The active form of FLP is short-lived. Thus, all of the induced events are likely to have occurred at or close to the time of heat shock.

4   By using a tissue-specific regulatory element (rather than a heat-shock regulator), one can control where in the fly the exchanges occur (rather than when). The value of this technique is that one can use intra-chromatid exchanges between FRT elements flanking a gene of interest to excise that gene in a specific tissue (figure 5.7).

Similar systems involving the cre–lox system have been developed extensively for use in mammals (Copeland et al. 2001) and in Arabidopsis (Hoff et al. 2001). These tools allow for the facile creation of clones at various times during development.

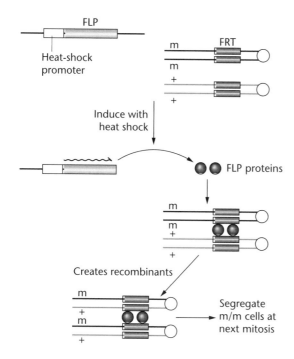

Figure 5.6 The FLP–FRT system

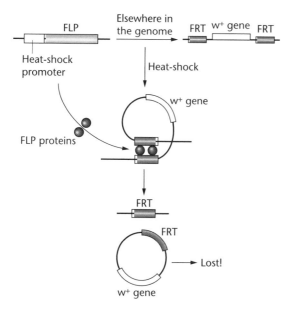

Figure 5.7 Use of the FLP–FRT system to excise a gene of interest

Summary

This chapter has addressed two different aspects of the same question: when and where do genes work? Section 5.1 dealt with the "when" in the context of biosynthetic pathways or regulatory hierarchies, and the "where" in the context of whether or not two or more acted in the same or different pathways. Section 5.2 asked the same questions in terms of the when and where of a developing organism. Both types of analysis can be extremely powerful if used correctly; however, they also require a thorough understanding of the nature of the mutants being used.

Genetic fine-structure analysis

To the early geneticists genes were discrete units of both function and structure. A hundred years later, we understand that genes have defined length and can be divided into separate functional domains. For example, genes are composed of regulatory regions, promoters, 5′ untranslated regions, introns/exons, and 3′ untranslated regions. The proteins they encode are often comprised of a number of functionally separate domains, each of which executes a different function. It is precisely this complexity of gene structure that allows us to obtain the occasional partial loss-of-function mutants that destroy the function of only one of those multiple domains. The question then becomes: can we use such mutants to help identify structure and structural significance in a given gene or its protein product? To address that question, we must first map our mutations within the gene. We will begin with a historical discussion of the discovery of intragenic recombination.

6.1   Intragenic mapping (then)

6.1.1   The first efforts towards finding structure within a gene

As noted above, our early intellectual ancestors viewed the gene as an indivisible particle that comprised the fundamental units of mutation, recombination, and function. Looking back at these early concepts of the gene, they seem hard to imagine. Our students often ask "didn't they realize that genes are made out of DNA, and that DNA has length?" No, they did not. Remember, these early geneticists worked before Watson and Crick (1953) solved the structure of DNA. More critically, no one had performed the right experiment to challenge the view that genes were particulate in nature.

Such experiments were first done by Pete Oliver and Mel Green (Green 1990) using alleles at the *lozenge* (*lz*) gene on the X chromosome of Drosophila. Like many great experiments, the result was obtained as a consequence of a carefully performed experiment done for another purpose. Mutations at the *lozenge* gene are notable for their effects on the shape and texture of the Drosophila eye. Many *lozenge* alleles are also female sterile. However, some *lz* alleles, or combinations of *lz* alleles, do allow the production of a few progeny (Oliver 1940; Oliver & Green 1944). Oliver and Green were examining the fertility of various double-heterozygote combinations of *lz* alleles in an attempt to study interactions between various *lozenge* alleles and their effect on female fertility. Much to their surprise, *lz+*, or wildtype, male progeny emerged from several types of double heterozygotes. Such progeny were quite rare, approximately one per 1,000, but the experiment was clearly reproducible.

Three lines of evidence argued that these *lz[+]* male progeny did not arise by simple reversion of one of the two *lz* alleles. First, wildtype progeny were not produced by *homozygotes* for the semifertile alleles (i.e. *lz[x]/lz[x]*). Second, Mel Green used a clever set of genetic tricks to demonstrate that the reciprocal product (the double-mutant chromosome *lz[x] lz[y]*) was also produced (Green & Green 1949; Green 1990). Third, production of *lz[+]* or double-mutant combinations was often associated with the recombination of flanking markers. Thus for the genotype *A lz[1] B/a lz[2] b*, where A and B are flanking genes:

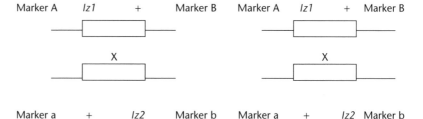

the *lz+* wildtype recombinant products were usually *aB*, while the double mutant (*lz[1] lz[2]*)-bearing chromosomes were *Ab*. The only reasonable explanation for these data is that both types of chromosomes arose by recombination. A similar observation for another locus was obtained by Professor Ed Lewis at Cal-Tech.

These experiments demonstrate that the gene is divisible by recombination. The importance of this result cannot be overstated. As noted by Pontecorvo in 1958, "The most obvious wrong idea is that of the particulate gene, i.e., of the genetic material as beads on a string in which each bead is the ultimate unit of crossing over, of mutation and of (function)." If recombination can occur between two different mutations within a gene, then the gene must have length and the two mutations must have occurred at different sites within the gene. Not only is the gene no longer the unit of recombination, it is no longer the unit of mutation. Different mutations cannot be thought of as different quantum states of an atom-like gene, but must be viewed as different lesions along a linear gene.

Still, Mendel's view of the gene as a particle was difficult to dislodge, despite the existence of the necessary evidence. Change would be created by a rather brash set of young biologists who viewed the world not through the Mendelian inheritance of flies and corn, but rather through the most unusual genetics of bacteria and bacterial viruses. One of the most influential of these geneticists was Seymour Benzer. Beginning his genetic studies after the elucidation of the structure of DNA, Benzer was able to dissect the anatomy of the gene.

6.1.2   The unit of recombination and mutation is the base pair

Following the discovery that DNA was the hereditary material (Avery et al. 1944) and the elucidation of its structure (Watson & Crick 1953), the concept of a linear gene with many mutable sites seemed quite reasonable. The elegant work by Seymour Benzer on the phage T4 allowed two adjacent genes to be dissected in a fashion that identified hundreds of separately mutable sites within each gene, each of which could be recombinationally separated from the others (see below). Benzer was able to accomplish this feat both because of the ease with which phage recombination studies can be performed, and because of his cleverness in choosing mutants that were amenable to selecting wildtype recombinants.

Benzer (1955, 1962) exploited the discovery (by others) of *rII* mutants in phage T4. Wildtype T4 produces plaques with a normal appearance on three different *E. coli* strains: S, B, and K. The ability of wildtype and *rII* mutant T4 phages to produce plaques on these three strains is listed in the table below. (NOTE: A plaque is a clear region produced on the surface of a culture in a glass dish where phage particles have multiplied and destroyed the bacterial cells.)

	Strain S	Strain B	Strain K
Wildtype	Normal plaque	Normal plaque	Normal plaque
rII	Normal plaque	Large, round plaque	No plaque

Understanding the value of these three strains is critical to understanding Benzer's analysis. The unusual plaque phenotype on Strain B allows *rII* mutants to be easily isolated. (Other types of mutants that produce large round plaques will also be isolated by this type of screen, but only *rII* mutants fail to produce any plaques on Strain K.) The inability of *rII* mutants to produce plaques on Strain K allows the easy identification and collection of wildtype revertants or recombinants. Finally, under conditions where it is important to easily grow *rII* mutants in a non-selective fashion, one can use Strain S.

Benzer identified *rII* → wildtype revertants by first growing *rII* mutant phage in Strain S, and then plating the phage on Strain K. The various *rII* mutants tested showed a wide range of revertant frequencies, ranging from one in 10^3 progeny to less than one in 10^8 progeny phage. These mutants were presumed to be single base changes, or point mutants, that reverted the original mutants.

Crosses of different *rII* alleles often produced wildtype recombinants at a frequency of a few percent. To obtain these recombinants, and measure their frequency, Strain B cells are co-infected with mutant *rII*(X)-bearing phage and mutant *rII*(Y)-bearing phage. The resulting lysate is then plated on Strain K and on Strain B. The number of wildtype phage produced by recombination can be determined by the number of plaques on Strain K. The total number of plaques on Strain B is an estimate of the total progeny. The ratio of Strain K/Strain B plaques is an estimate of the recombination frequency. Using such crosses, Benzer was able to create a linear map for eight spontaneously arising *rII* mutants.

The discovery of non-reverting *rII* mutations, corresponding to deletions, enhanced the ease of mapping *rII* mutants (i.e. no revertants of these mutations were observed in more than 10^{10} progeny phage). In an important step towards fine-structure mapping, Benzer mapped these deletions relative to each other by determining which deletion-bearing phage could recombine to form a wildtype phage. To do this, Strain B cells are co-infected with *rII* deletion (X)-bearing phage and *rII* deletion (Y)-bearing phage. The resulting lysate is then plated on Strain K and on Strain B. Two overlapping deficiencies cannot recombine to produce a wildtype recombinant, but two non-overlapping deficiencies can recombine at some frequency to produce a wildtype. An example of a deficiency map and the recombination data obtained from pairwise crosses is presented below:

(A) Positions of the four deficiencies

Deletion

1

2

3

4

(B) Production of wildtype recombinants

+ = wildtype recombinants recovered
− = no wildtype recombinants recovered

		Parent Y			
		1	2	3	4
Parent X	1	−	−	+	+
	2		−	−	+
	3			−	−
	4				−

It should not surprise you that each deletion failed to recombine with a defined set of revertible mutants (point mutations). Point mutations will not be able to recombine to produce wildtype progeny with a deletion that removes the base altered by that mutant. In the diagram below, the dashed lines indicate the lengths of the two different deletions (A and B), and the symbol X denotes a point mutant:

Mutant 1 ——————————X——————————

Deletion A ———————— – – – – ————————

Mutant 1 ——————————X——————————

Deletion B ———————————————— – – –——

There is no way a phage-carrying mutant 1 and deletion A can recombine to create a wildtype phage, because there is no wildtype base pair corresponding to mutant X. However, a phage-carrying mutant 1 and deletion B can recombine to produce a wildtype because the wildtype base pair is present. By mapping deficiencies relative to each other, Benzer was able to divide the *rII* region of page T4 into 80 distinct subintervals.

Most deficiencies of the *rII* region failed to recombine with a number of revertible point mutants. Furthermore, these point mutants recombined with each other at a very low frequency. To explain these two phenomena, Benzer reasoned that all of these point mutants fell within one deletion. As shown by the diagrams below, one can easily map a number of point mutants by allowing them to recombine with a set of overlapping deficiencies:

(A) Map of six point mutants (A–F) and four deletions (1–4) in the *rII* region of T4

	A*	B*	C*	D*	E*	F*

Deletion

1 ————————

2   ——————————

3   ——————————————

4   ——————————

(B) Production of wildtype recombinants

+ = wildtype progeny recovered
− = no wildtype progeny recovered

Deletion	Mutant					
	A	B	C	D	E	F
1	−	−	+	+	+	+
2	+	−	−	+	+	+
3	+	+	−	−	−	+
4	+	+	+	+	−	−

Using this method, Benzer mapped over 1,000 *rII* point mutants within the 80 subintervals defined by the deletions. The point mutants within each subinterval could then be mapped against each other. Those pairs of point mutants that could recombine to produce a wildtype phage were said to define separate sites within the interval, while those mutants that could not recombine to produce a wildtype phage were said to define the same site. Based on this analysis, the *rII* region could be divided into approximately 350 separately mutable sites. Benzer argued, on statistical grounds, that at least 100 or more sites remained to be discovered, raising the number of separately mutable sites to at least 450. Most of these sites (~375) mapped in one complementation group (see below), which Benzer named the *rIIa* gene. The others mapped in a second complementation group, denoted *rIIb*. Benzer estimated that the *rIIa* and *rIIb* genes were each 4,000 base pairs in length. These base pair estimations would argue that the units of mutation could be no longer than approximately ten base pairs, and that recombination occurs in intervals of ten base pairs or less in length. (In fact, subsequent DNA sequencing of phage T4 would reveal that Benzer overestimated the physical lengths of these genes: *rIIa* is only 2,178 base pairs in length and *rIIb* is only 939 base pairs in length.)

The final demonstration that the unit of mutation was the base pair, and that the unit of recombination was the space between two adjacent base pairs, was accomplished by Charles Yanofsky and his collaborators. While working on the tryptophan synthetase gene in *E. coli*, Yanofsky identified mutations at each of three bases within a single codon (#211) that altered the amino acid specified by that codon. Specifically, these mutations altered the gene to produce the mRNA codons shown below:

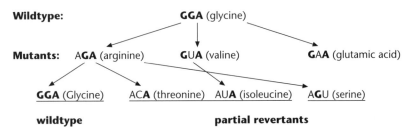

Wildtype: **GG**A (glycine)

Mutants: A**GA** (arginine)   **G**U**A** (valine)   **G**A**A** (glutamic acid)

GGA (Glycine)   AC**A** (threonine)   AU**A** (isoleucine)   A**G**U (serine)

wildtype   **partial revertants**

These data demonstrate that each base pair of the codon is separately mutable. More critically, Yanofsky demonstrated that two mutations, in separate but adjacent base pairs, could recombine to produce a wildtype tryptophan synthetase gene, but two mutations in the same base pair could not. For example, a bacterial cell carrying a copy of the AGA (arginine) codon and the GAA (glutamic acid) codon at position 211 could have a recombination event between the first two base pairs of this codon to yield the wildtype GGA (glycine) codon:

$$A\text{–}\mathbf{G}\text{–}\mathbf{A}$$
$$\times \qquad \longrightarrow \mathbf{G}\text{–}\mathbf{G}\text{–}\mathbf{A}$$
$$\mathbf{G}\text{–}A\text{–}A$$

Taken together, these studies performed by Benzer and Yanofsky reduced the unit of mutation and recombination from the "gene" to the base pair. But if the gene is not the unit of mutation or recombination, then just how do we define a "gene?" Benzer (1955, 1962) chose to address this issue by formalizing the concept of the gene as the unit of function. Benzer noted that co-infection of Strain K with wildtype and *rII* phages, followed by plating on Strain B, produced both mutant and normal plaques. (Remember, *rII* mutants cannot survive in Strain K.) The wildtype phage must provide the necessary function to the *rII* phage in *trans*, allowing the *rII* phage to replicate in the Strain K cell. When Strain K cells were co-infected with different pairs of *rII* mutants, two different results were obtained. For some combinations of alleles no progeny phage were produced (no lysis), while for some combinations billions of progeny phage were created (lysis).

These observations allowed Benzer to develop a test for gene function that he referred to as the *cis–trans* test. The formal basis of this test is described in figure 6.1. If two mutations define the same gene, then co-infection of an *m1* phage and an *m2* phage (the *trans* test) should fail to produce lysis in Strain K. Infection of a + + phage and an *m1 m2* phage should show good lysis in Strain K (the *cis* test). When two mutations are in different genes, the results of the *cis* and *trans* tests should be identical. (What Benzer has just described is, in fact, a rather rigorous variant of the standard complementation test that we described in some detail in chapter 3.)

Note that these experiments are quite different from the recombination experiments described above. In an experiment designed to assess the capacity of two mutants to produce wildtype phage by recombination, the two viruses were co-infected in a semipermissive (Strain B) or permissive (Strain S) host, and the presence of wildtype recombinants detected by plating the resulting progeny phage on Strain K. In this case, where we are assessing function, phage bearing the two different *rII* alleles are co-infected into the highly restrictive (Strain K) host. Unless the two phage can themselves provide the necessary lytic functions without the need to produce a recombinant, infection will virtually always be abortive.

Benzer observed that the revertible point mutants could be divided into two groups (*rIIa* and *rIIb*) by this experiment. Co-infection of Strain K with either two *rIIa* alleles or two *rIIb* alleles produced no progeny (because both phage were deficient for the same function). But co-infection with an *rIIa* and an *rIIb* allele did produce phage. Quoting Benzer, "If both mutants belong to the same

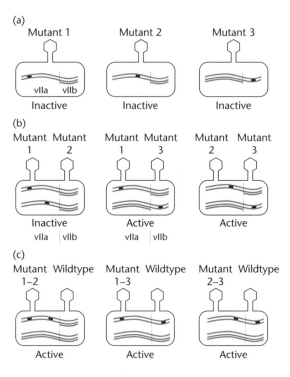

Figure 6.1 The *cis–trans* test determines whether or not two mutants map within the same gene. When two different T4 phages (both carrying an *rII* mutant) infect the same bacterial cell from Strain K, the infection will only produce phage progeny in the two mutants that lie in the two different cistrons (genes) *rIIa* and *rIIb*. A test of each mutant with the wildtype phage provides an important control. (Adapted from Benzer 1962)

segment, mixed infection on K gives the mutant phenotype (very few cells lyse). If the two mutants belong to different segments, extensive lysis occurs with liberation of both infecting types (and recombinants). Thus, on the basis of this test, the two segments of *rII* correspond to independent functional units." Fortunately, the *rIIa* mutants mapped on one side of Benzer's *rII* map, while the *rIIb* mutants mapped on the other. These results divided the *rII* locus into two adjacent but functionally separate segments, *rIIa* and *rIIb*.

Benzer redefined the concept of the gene (or *cistron* as he called it) as *being* the unit of function. He further showed that each of these functional units could be further divided into a number of separately mutable sites, each of which was separable by recombination.

6.2   Intragenic mapping (now)

The pioneering studies of Green and Oliver emboldened eukaryotic geneticists to attempt to fine-structure map eukaryotic genes. However, intragenic recombination frequencies are often quite low in higher organisms, and strong selection techniques had to be applied to recover recombinants. We will not discuss the details of any of these studies for two reasons. First, geneticists no longer position mutants by intragenic recombination, and have not done so for more than a decade. It is more straightforward and efficient to sequence even a large collection of mutants. Second, the analysis of the process of recombination itself using intragenic recombination is considered in the next chapter. But the process of aligning the map with complex complementation patterns, or a diversity of phenotypes exhibited by mutants in a given gene, is a problem that still very much vexes the modern geneticist. It is this problem that we shall consider in detail below.

[For those students curious enough to want to see how such intragenic maps were constructed prior to the availability of DNA sequencing, the following papers might prove useful. Methods for the construction of genetic fine-structure maps in Drosophila are reviewed by Finnerty (1976), Judd (1976), and Chovnick (1989). McKim et al. (1988a,b), Rand (1989), and Ruvkun et al. (1989) have described fine-structure mapping in *C. elegans*. Finally, intragenic recombination in maize has also been studied at several loci in maize, and the paper by Okagaki and Weil (1997) provides an excellent introduction to this literature.]

6.3 Intragenic complementation meets intragenic recombination: the basis of fine-structure analysis

Genetic fine-structure analysis combining intragenic mapping with intragenic complementation studies has been done in a large number of eukaryotes. These studies have allowed investigators to observe three primary types of intragenic complementation, and to use those findings to deduce various aspects of gene structure and function. The first class of interactions involves those genes that encode polyfunctional proteins. It is often possible to obtain mutants in such genes that disrupt only one of the two or more functional domains of proteins. Two such mutants can display intragenic complementation if they fall in different domains. They will obviously not be able to complement each other if they fall in the same domain. We can thus group the mutants together on the basis of their pattern of intragenic complementation. In the case where we can map mutants that ablate the same function to a specific subregion of the gene's map, we can correlate position within a gene with a specific function of the protein product.

The second two classes of intragenic complementation include cases in which one, or both, of the mutations define transcriptional regulatory elements. For example, if a given gene possesses both wing- and leg-specific enhancers, a mutation in the leg-specific enhancer (which still allows the gene to function in the wing) might well complement a mutation in the wing-specific enhancer (which still allows the gene to function in the leg). The two classes of this type of interaction are differentiated by their sensitivity to chromosome aberrations that disrupt somatic pairing of the two alleles. In the case just presented, complementation does not require gene pairing, but there are notable examples in which this type of complementation is exquisitely pairing-dependent. These cases, which reflect the ability of regulatory elements to act in *trans* on paired homologs, are referred to as examples of **transvection**. Cases of transvection usually involve one enhancer mutation paired with a mutation in the allelic structural gene.

We begin our discussion with the first class of intragenic complementation, the dissection of genes encoding polyfunctional proteins. After laying a bit of theoretical groundwork, we present the example of the HIS4 gene in baker's yeast. Box 6.1 at the end of the chapter describes the historical "gory details" of a more complex, but conceptually quite similar, analysis of the *rudimentary* gene in Drosophila. In both of these cases, the success of the analysis depended on multiple factors: (i) the ability to construct a high-resolution fine-structure map of the gene that positioned a large number of mutants in an unambiguous order; (ii) the existence of point mutants that displayed intragenic complementation; and (iii) the existence of tools that allowed one to subdivide the function of the gene product, so that components of protein function and components of gene structure could be connected.

6.3.1   *The formal analysis of intragenic complementation*

In order to analyze intragenic complementation maps we need to apply a permutation of the complementation test that was not discussed in chapter 3. Let us suppose that gene A encodes a protein with three functionally and spatially discrete functions: X, Y, and Z. We can get missense mutants that specifically knock out one of these functions without affecting the other two. Each of these sets of mutants might appear to form a single complementation group, and members of that group would be expected to complement similar function-specific mutants in the other separate elements of the gene. However, we also expect mutants (such as a frameshift or early nonsense mutant) that knock out the entire gene to form a large complementation group. The members of this "fully non-complementing" group are expected to fail to complement all of the function-specific mutants. How then do we combine these data to identify separate functional units within a set of what appear to be overlapping complementation groups?

We need to distinguish between **complementation units** and **complementation groups**. We state that a given set of mutants define **complementation group A** as long as *all* of those mutants fail to complement each other in transheterozygotes. Now suppose that there are two other complementation groups (**B** and **C**) such that mutants in **B** fully complement the mutants in group **A**, but mutants in **C** fail to complement both **A** and **B** mutants. We can diagram the complementation pattern as follows:

Group A	Group B
A1, A2, A3 .... An	B1, B2, B3, B4 .... Bn

Group C

C1, C2, C3, C4 .... Cn

_____| Gene A |_____

We can now subdivide this map into two **complementation units**, **unit I** and **unit II**. The rule here is that those mutants capable of intragenic complementation in unit I will fail to complement each other, but they may complement the complementing mutants in other units. Some mutants may obviously define more than one unit.

If we keep isolating more mutants in this gene (our reference point for "in this gene" becomes a failure to complement the "fully non-complementing mutants," i.e. the mutants that fail to complement *all* other mutants), we might be able to build a rather more complex map:

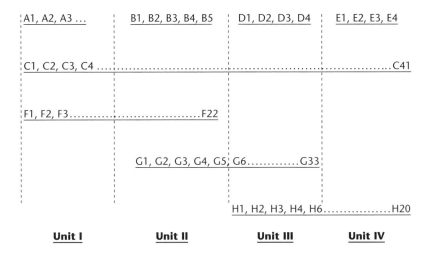

In order to proceed with this analysis, we need to have two further conditions met:

1 The order of mutants, as revealed by intragenic mapping studies, needs to be consistent with the order of complementation units. (This is simply a restatement of the idea that there are functionally and spatially separate functional domains. If such domains exist, they can be mapped, and if they can be mapped, then they probably have functional significance.)

2 The mutants in each of the units can be correlated with specific and different functional defects. For example, this gene B makes a protein with four enzymatic activities (**W**, **X**, **Y**, **Z**) such that mutants in unit I ablate activity **W**, mutants in unit II ablate activity **X**, mutants in unit III ablate activity **Y**, and so on.

6.4   An example of fine-structure analysis for a eukaryotic gene encoding a multifunctional protein

6.4.1   *A genetic and functional dissection of the* HIS4 *gene in yeast*

In the 1960s and 1970s, Gerry Fink characterized the gene–enzyme relationships for histidine biosynthesis in yeast. These studies clearly showed that most of the genes involved in this process were widely dispersed in yeast. However, mutants at the *HIS4* locus affected one, two, or three enzymes involved in this process, namely: phosphoribosyl-AMP cyclohydrolase, phosphoribosyl-ATP pyrophosphatase, and histidinol dehydrogenase. Phosphoribosyl-AMP cyclohydrolase and phosphoribosyl-ATP pyrophosphatase catalyse the second and third steps in histidine biosynthesis, respectively, and histidinol dehydrogenase catalyses the last two steps.

To further study this locus, Fink analyzed 58 new *his4* mutants in terms of their intragenic complementation patterns and their capacity to produce the various enzymatic activities encoded by *his4* (Fink 1966) The positions of the mutants were mapped by intragenic recombination and all 3,364 pairwise complementation tests were performed. As shown in figure 6.2, these tests reveal three separate complementation units, each of which is uniquely associated with one of the three enzymatic defects. Thus, Fink's data also suggest that the three enzymatic activities were the result of the polar translation of a single mRNA molecule. To prove this he demonstrated that all of his tRNA mutant-suppressible

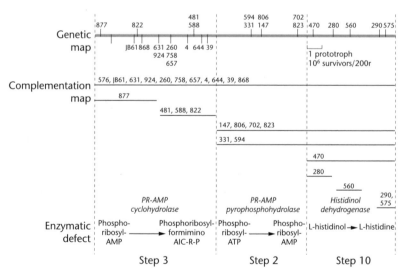

Figure 6.2   Fink's complementation map of the *HIS4* gene in yeast. The enzymatic defects for mutations in each region are indicated at the bottom of the figure. (After Fink 1966)

nonsense alleles fell into the following classes: entirely non-complementing A–B–C– mutants, B–C– mutants, and a C– mutant. These nonsense mutants show a polarized loss of activity, similar to the polar mutants observed for the Lac Operon in *E. coli*. Indeed, Keesey et al. (1979) would go on to show that the product of the *his4* gene in yeast is a trifunctional protein of ~95 kD.

Box 6.1 describes a similar analysis of the *rudimentary* gene in Drosophila. The ability of this type of genetic fine-structure analysis to identify discrete and separable functional domains within a single gene is now a well-documented and frequently used tool in eukaryotic genetics [for recent examples, see Gepner et al. (1996), Tang et al. (1998), and Ponomareff et al. (2001)]. The primary difference in modern studies, compared to their historical counterparts, is that mutants are quickly mapped by sequencing and putative functional domains can often be predicted by computer searches.

6.5   Fine-structure analysis of genes with complex regulatory elements in eukaryotes

In the examples described below, we will focus on genes with functionally independent regulatory elements that control gene expression in different tissues, or at various times during development. In the previous cases, we considered mutants that failed to complement all other mutants at a locus (i.e. those that did not display intragenic complementation) as mutants that ablate protein production. We considered the intragenic-complementing alleles to be missense mutants that disrupt one of several separable enzymatic activities within the same protein. For genes with complex regulatory elements, our null alleles will usually define the transcription unit, while alleles retaining some degree of function will define one of many upstream regulatory elements. (Be aware of a critical component of this analysis: *the assumption that genetic regulatory elements act only in cis*. However, also be aware that *trans*-acting gene activation is well known in Drosophila. We will consider *trans*-acting cases in the second example.)

6.5.1   *Genetic and functional dissection of the* cut *gene in Drosophila*

The *cut* gene in Drosophila was originally named for alleles that caused the loss of the wing tips, resembling cutting by scissors. However, most of the viable *cut* alleles have other phenotypic defects such as kinked legs (kinked femur), missing or reduced bristles, and aberrations of head or eye shape. There are also lethal alleles of this gene. The lethal alleles can be divided into three groups (*I, II, III*) based on the time of death (larval or embryonic) and their intragenic complementation patterns with respect to each other, and with respect to the viable alleles (Johnson & Judd 1979). A map of these alleles, created by superimposing intragenic mapping on a physical map, is presented in figure 6.3.

By looking at the intragenic complementation pattern in table 6.1, one can visualize the complexity of interactions at this locus (Liu et al. 1991). First, as noted by Jack and Judd (1979), the lethal alleles seem to map to the right side of

Chapter 6

Table 6.1 Complementation of *cut* locus mutations. (After Liu et al. 1991)

	kf	ct	L-39	L-59	1S2	1188	1221	1242	L-1	L-5	C145	1S
			Lethal III		Lethal IV				Lethal I		Lethal II	
kf	+	+	+	+	+	+	+	+	+	+	kf	kf
ct	+	+	ct	ct	+	+	+	+	ct	ct	ct	ct
L-39			—									
L-59			0 (300)	—								
1S2			0.23 (210)	0.05 (311)	—							
1188			0.95 (395)	0.12 (427)	0.78 (198)	—						
1221			0.23 (244)	0.35 (171)	0.02 (183)	0.18 (219)	—					
1242			1.3 (443)	0.61 (298)	0.46 (245)	0.92 (339)	0.55 (406)	—				
L-1			0.01 (148)	0 (153)	0 (63)	0.99 (279)	0 (115)	0.52 (265)	—			
L-5			0.08 (155)	0 (290)	0 (75)	0.17 (327)	0.13 (146)	0.09 (348)	ND	—		
C145			ND	0 (156)	0 (177)	0.02 (213)	0 (189)	0 (405)	ND	0 (50)	—	
1S			ND	0 (134)	0.006 (156)	0 (236)	0 (199)	0.003 (343)	0 (320)	0 (334)	0 (35)	—

kf, kinked femur phenotype; ct, cut wing phenotype; ND, not done. The numbers outside parentheses indicate the fraction (number of heterozygous mutants surviving)/(number of wildtype controls). This is the fraction of what is expected if the mutations fully complement. The number in parentheses is the number of control flies.

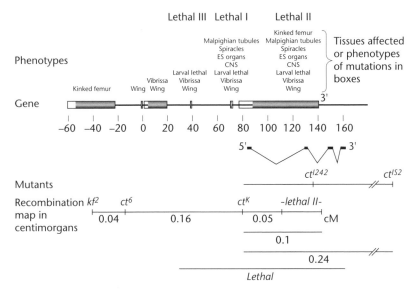

Figure 6.3 A map of the *cut* gene in Drosophila. (After Liu et al. 1991)

the locus while the viable (visible) mutants seem to map to the left side of the locus. The cloning of the *cut* locus (Jack 1985; Blochlinger et al. 1988) revealed that the transcribed portion of the gene corresponded to lethal groups I and II (i.e. to the "right side" of the gene). The cloning also revealed that the leftward 140 kb of the locus is comprised of the visible mutations. In other words, the mutants that map downstream of the transcriptional start are functionally null, blocking gene expression in all tissues giving rise to the lethal phenotype. However, the mutants that map upstream of the transcription start block expression in a subset of tissues, giving rise to the other visible phenotypes.

Most of the upstream mutations are the result of insertions by the retrotransposon *gypsy* (Jack 1985). As pointed out by Jack and DeLotto (1995), "these gypsy insertions are polar in nature, the ones nearest the promoter block expression in the most tissues and those farthest away affect the fewest." All of the *gypsy* mutations, even those 140 kb from the cut gene's promoter, cause the kinked femur/cut wing phenotype. Mutants at an intermediate distance can cause larval lethality, without affecting cut gene expression in the embryo. Again quoting Jack and DeLotto (1995), "Thus, *cut gypsy* insertion mutants accumulate phenotypes as they get closer to the promoter."

Prior to these studies, *gypsy* insertion elements had been shown to block gene expression when inserted between an enhancer element and its target promoter (Geyer et al. 1986; Geyer & Corces 1987; Harrison et al. 1989). As discussed previously, this blocking requires the binding of the SuHw protein to this *gypsy* element. Finally, Jack and DeLotto (1995) went on to show that the upstream visible mutants *do* define enhancer-like regulatory elements with the expected tissue-specificities.

6.6   Pairing-dependent intragenic complementation

The type of intragenic complementation seen for the *cut* locus is commonly observed in eukaryotes. However, in Drosophila this type of complementation often takes on one more degree of complexity. Perhaps because somatic chromosome pairing seems to be ubiquitous in Drosophila, enhancer elements often have the capacity to function not only in *cis*, but also on their homolog in *trans*. This allows for some rather curious examples of intragenic (and interallelic) interactions and complementation, as described below.

6.6.1   *Genetic and functional dissection of the* yellow *gene in Drosophila*

The *yellow* (*y*) locus of Drosophila produces a pigment that darkens the color of the wings, body, bristles, and tarsal claws of the fly. Tissue-specific enhancers required for expression of yellow in the bristles and tarsal claws are located in the single intron of the *yellow* gene, while the body and wing enhancers are located in the upstream 5-regulatory region. Null alleles of yellow affect expression in all four tissues, presumably by preventing protein production. However, there are mutants that are specifically defective in either wing or body pigmentation alone. Not surprisingly, reciprocal pairs of such mutants can display intragenic complementation to produce a wildtype fly. The basic principles of this interaction are identical to those described above for the *cut* gene.

Curiously enough, there exist several recessive yellow mutants which lack both pigmentation in the wings and the body, but which can complement each other to produce a fly with wildtype pigmentation. As shown by Geyer et al. (1990), this interaction requires that the two mutants be aligned with each other in identical positions on homologous chromosomes (recall our discussion on transvection in chapter 3). Consider the interactions observed between the *y[2]* and the *y[1#8]* mutants diagrammed in figure 6.4. The *y[2]* mutant results from a gypsy insertion between the wing–body enhancers and the promoters, which occludes the functions of these enhancers. As a result, the mutant gene only produces pigment in the bristles and tarsal claws. The *y[1#8]* mutant is the result of a 0.8 kb deletion that removes the yellow gene promoter, creating a null mutant. However, near normal wing and body pigmentation is observed in *y[2]/y[1#8]* transheterozygotes. As pointed out by Morris et al. (1999a), "this interaction can be explained by the action of the wing and body enhancers of *y[1#8]* in trans on the *y[2]* promoter when the two alleles are in close proximity."

This interaction is more complex than a simple "enhancers can sometimes act in *trans*" model might suggest. Data obtained by Wu and her collaborators (Morris et al. 1998, 1999a,b) argues that the enhancers at *yellow* will only act in *trans* if the *cis* promoter element has been crippled or removed. All *yellow* alleles that support transvection with *y[2]* carry a deletion of or insertion into the promoter region. Morris et al. (1999a,b) present a variety of molecular explanations for this phenomenon.

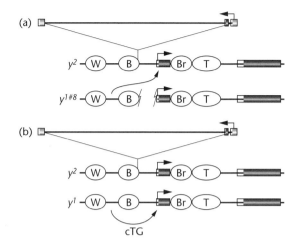

Figure 6.4 Transvection at the *yellow* gene in Drosophila. The large insertion into the *y[1#8]* allele is *gypsy* element (with Su(Hw) binding sites) that is transcribed in the opposite direction of the *yellow* gene. The symbols W, B, Br, and T indicate enhancer elements. The W and B enhancers can function in *trans*, as shown in (a), if their *cis* promoter is non-functional. However, if, as in (b), the *cis*-promoter is functional, these enhancers cannot function in *trans*. (After Morris et al. 1999a)

In some cases intragenic complementation may be sensitive to pairing, causing a complex pattern. Another complex example of pairing leading to observed complementation patterns is the interaction of the *white* and *zeste* genes in Drosophila.

6.6.2 The influence of the zeste gene on pairing-dependent complementation at the white locus in Drosophila

Mutants at the *white* locus in Drosophila affect the color of the fly's eyes as well as pigmentation in several other tissues. Null alleles of the *white* gene produce a bright white eye phenotype. Meanwhile other alleles produce intermediate levels of pigmentation (e.g. white cherry, white apricot, white eosin, etc.). Intragenic mapping studies by numerous investigators allowed the construction of the map shown in figure 6.5.

As indicated in figure 6.5, the various white alleles differ from each other in three ways, other than pigmentation. First, some of the mutants exhibit proper dosage compensation, while others do not. Second, one mutant *w[sp]* affects the pattern of pigment distribution, while the others affect the amount of pigment deposited. Finally, and perhaps most curiously, the mutants differ in terms of their interaction with *zeste* (*z*). Mutants at the *zeste* locus can diminish the activity of the *white* gene, but only when the two homologous *white* genes are paired. The *z w+/z w+* females have a lemon color eye, while the *z w+/Y* males have wild-type eyes. This effect is not sex-limited, *z w+/z Df(w+)* females also have normal

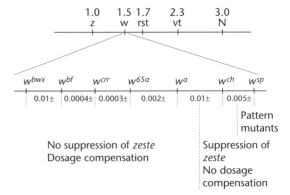

Figure 6.5 A map of the *white* gene in Drosophila showing the positions of seven mutant sites separable by recombination

eye color, and males who carry a zeste mutant and a tandem duplication of w+ have lemon-colored eyes.

Some *white* mutants, such as *white-cherry* (w^{ch}) and *white-65a* (w^{65a}) act as suppressors of the *white–zeste* interaction, as does the deficiency for the white locus. A *z w[ch]/z w[+]* female has wildtype eyes because the *zeste–white* interaction is suppressed, while a *z w[65a]/z w[+]* female has lemon eyes. Mutants that map to the right of the white locus that suppress the *zeste–white* interaction and fail to show dosage compensation (elevated transcription in males) have been shown to define the 5′ regulatory regions of the *white* gene. Mutants that map to the left of the white locus define the coding region. The *zeste* gene has been shown to encode a transcription factor that binds to regulatory elements of the *white* gene.

The interaction of the Zeste[1] protein with the *white* gene requires pairing of two *white* alleles. The mechanism of this interaction remains unclear [cf. Rosen et al. (1998)]. It appears that the wildtype function of the Zeste protein is to activate *white* gene transcription and to mediate inter-chromosomal interactions between the two *white* genes. The Zeste protein produced by the canonical *z[1]* mutation still interacts with paired *white* genes, but represses *white* gene expression (Rosen et al. 1998).

Of perhaps more interest to us is the finding that *zeste* also mediates intragenic complementation at the *white* locus. Babu and Bhat (1980) demonstrated that although most *white* alleles fail to complement each other, the rightmost mutation (*w[sp]*) complements virtually all other *white* alleles, except in the presence of at least one allele of *zeste* (*z[a]*). The lack of a functional Zeste protein precludes intragenic complementation. We will show below that the *z(a)* mutant also prevents pairing-dependent complementation at two other loci (*BX-C* and *dpp*).

The zeste–white interaction is at least in some ways similar to the *yellow* gene, in that intragenic complementation is sensitive to pairing interactions. The zeste–white example is more complex in the sense that the zest protein is needed to facilitate the pairing interaction. The ultimate example of proteins mediating pairing interactions is the BX-C complex of Drosophila.

6.6.3   Genetic and functional dissection of BX-C in Drosophila

The structure of the bithorax complex (BX-C) in flies was determined by Ed Lewis. This locus controls the development of segment identities for most of the thorax and for the abdomen of the developing fruit fly. For this purpose, we will use a simplified genetic map drawn below:

bx   Cbx   Ubx   bxd   pbx

The locus is defined by recessive and dominant alleles that exhibit separate phenotypic effects on development and can be separated by intragenic recombination. The *bx* mutants cause the anterior portion of the metathorax (AMT) to develop as the anterior mesothorax (AMS) normally would. Similarly, the *pbx* mutants cause the posterior portion of the metathorax (PMT) to develop as the posterior mesothorax (PMS) normally would. Combine these two mutants as a double homozygote and you get a fly with two mesothoracic regions (i.e. with four wings.). The *bxd* mutants affect the development of the first abdominal segment.

Ubx mutants cause all three of these transformations to occur, although the effect is weak in heterozygotes. Most Ubx alleles are recessive lethal, but fail to complement bx, pbx, and bxd alleles for their respective defects. The dominant Cbx mutant causes transformations that are the reverse of those caused by the bx and pbx mutants (i.e. mesothorax → metathorax).

The intragenic complementation patterns at this locus are complex. For example, a weak pbx-like PMT → PMS transformation phenotype is observed in *bxd +/+ pbx* heterozygotes, but none is observed in *+ +/bxd pbx* heterozygotes. *Ubx +/+ bxd* shows an extreme bxd-like phenotype, but the corresponding *Ubx pbx/+ +* double heterozygotes are nearly wildtype. Similarly, *Ubx +/+ pbx* shows an extreme pbx-like phenotype, but the corresponding *Ubx pbx/+ +* double heterozygotes are nearly wildtype.

Many of these interallelic interactions are sensitive to heterozygosity for chromosome rearrangements. For example, structurally normal *bx +/Cbx +* heterozygotes show only a weak bx-like transformation. In the presence of a large structural aberration on the same chromosome arm, a strong intensification of the bx-like transformation occurs. No such effect is observed in *cis*-double heterozygotes or in the presence of homozygosity for the rearrangement. Heterozygosity for a translocation or inversion breakpoint half a chromosome arm away can interfere with intragenic complementation at the BX-C. This effect is so strong that Lewis used it as a method to detect chromosomal aberrations as a means of biodosimetry for nuclear weapons testing.

Clearly, somatic pairing plays a critical role in the ability of the two BX-C alleles to interact, and these pairings can be disrupted by heterozygosity for chromosome rearrangements. Long-range transvection effects have also been documented at the *dpp* locus by Gelbart and his collaborators [cf. Gelbart (1982); Smolik-Utlaut and Gelbart (1987)].

Chromosome aberrations can also disrupt the pairing-dependent interactions at the *white* locus. However, the ability of aberrations to disrupt the zeste–white

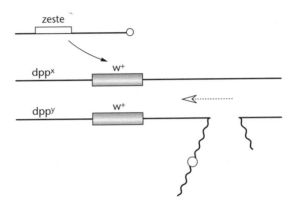

Figure 6.6   The Smolik-Utlaut and Gelbart experiment

interaction is inhibited only by aberrations with a breakpoint immediately adjacent to the white gene (Smolik-Utlaut & Gelbart 1987). Based on proximity-dependent differences in rearrangements, Smolik and Utlaut (1987) have argued that "the zeste–white interaction and transvection are two different proximity-dependent phenomena." This view is supported by an elegant experiment (figure 6.6) in which Smolik-Utlaut and Gelbart interposed paired *white* mini-genes between a heterozygous breakpoint and the *dpp* locus. The same breakpoint heterozygosity that prevented intragenic complementation at the *dpp* locus did not disrupt the ability of zeste mutants to interact with the paired *white* genes.

Golic and Golic (1996) have proposed an alternative interpretation of this result. Their view is that differing cell cycle times of relevant tissues may influence the ability of homologous chromosome arms to pair, even in the presence of structural heterozygosity. If that window of time, or "pairing interval," in the cells of the developing embryo (where the BX-C genes act) is shorter than the window for pairing in chromosome eye cells, then one could understand the differences in phenotype and proximity dependence. Quoting from Golic and Golic (1996), "Cells with a longer cell cycle have more time to establish the normal pairing relationships that have been disturbed by rearrangements . . . Minute mutations, which slow the rate of cell division, partially restore a transvection effect that is disrupted by inversion heterozygosity." We favor the Golics' interpretation of this result, and thus do not see any need to view the types of transvection observed at BX-C and white as different phenomena. Indeed, transvection at both BX-C and *dpp* are sensitive to the genotype of the fly at *zeste*, as well as for numerous other loci in the genome.

Summary

We began this chapter by describing the efforts to find structure within the gene. Our interest was in defining the unit of recombination and mutation, the base pair. But with the development of the *cis–trans* test, Benzer opened the door to finding functional structure within a gene. We then asked whether we could use a comparison of intragenic complementation maps to intragenic mutant maps to deduce which regions of a given gene executed which of two or more dissectable functions of that gene's protein product. As interesting as such methods are, we note that mapping mutants by intragenic complementation is decidedly outdated. One now maps mutants by sequencing, and functional domains of proteins are discovered in BLAST and motif searches. Still, out of these studies came the ability to sometimes dissect mutants that defined regulatory elements from those that defined structural genes, along with some surprising observations about the role of somatic pairing in gene function, at least in Drosophila, and perhaps in other organisms as well (Tartof & Henikoff 1991). There are still a few things about gene structure that one can only learn from "doing the genetics."

Box 6.1   Genetic and functional dissection of the *rudimentary* gene in Drosophila

Mutants at the rudimentary (*r*) gene in flies produce small truncated wings, sparse bristles, reduced fertility, and reduced viability. Early attempts to perform complementation tests and intragenic complementation analysis on the *r* locus (Green 1963) revealed a complex complementation pattern and three recombinationally separable sites. A detailed study (Carlson 1971) revealed an even more complex pattern. Although the 45 alleles could be aligned in a linear array (figure B6.1b), the mutants defined 16 complementation groups that only roughly correlated with map position (figure B6.1a). The complementation groups define seven complementation units within the *r* gene. The majority of complementation groups span more than one complementation unit. Indeed, as pointed out by Judd (1976), "less than half (18/45) of the mutants studied by Carlson behave as if they belong to only one of the complementation units."

For example, all three mutants in group #6 (30, 39, 40) fail to complement each other, and thus define a group. However, these three alleles also fail to complement the sole mutant (36) in group #1 and the mutant (29) in group #2, both of which fully complement each other. The single mutant (33) defining group #10 fails to complement all of the mutants in groups 1–3, while the mutant (31) defining group #8 fails to complement all of the mutants in groups 2–4:

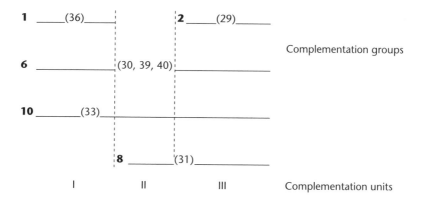

Box 6.1   (*cont'd*)

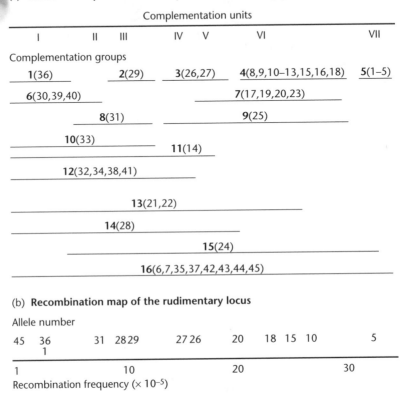

(a) **Carlson's complementation map of the rudimentary gene**

Complementation units

I	II	III	IV	V	VI	VII

Complementation groups

1(36)   2(29)   3(26,27)   4(8,9,10–13,15,16,18)   5(1–5)

6(30,39,40)   7(17,19,20,23)

8(31)   9(25)

10(33)

11(14)

12(32,34,38,41)

13(21,22)

14(28)

15(24)

16(6,7,35,37,42,43,44,45)

(b) **Recombination map of the rudimentary locus**

Allele number

45	36		31	28 29		27 26		20		18	15	10		5
	1													

1   10   20   30

Recombination frequency ($\times 10^{-5}$)

Figure B6.1   Carlson's complementation map of the *rudimentary* gene

These data defined both five complementation groups and three complementation units. But unit II is not defined by any mutants unique to that group, as are units I and III. Rather, its existence was inferred by the ability of unit I and unit III mutants to fully complement each other, while failing to complement mutants in the overlapping units. The key observation is that the complementation units seem to be collinear with the position of the mutants, as if the units correspond to discrete functional domains within the gene. Look at the single complementation group (#16) at the bottom of figure B6.1a. The mutants in this group fail to complement all of the other mutants in the rudimentary gene. Six of these eight non-complementing alleles map to the left end of the gene, suggesting a critical regulatory region.

One possible explanation for the complexity of these complementation data was the complexity of the collection of *rudimentary* mutants. Carlson used a large number of mutants created by a diverse set of investigators using different methods. When Rawls and Porter (1979) repeated these studies using a new set of mutants, only four

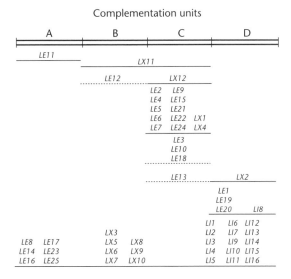

Figure B6.2 The Rawls and Porter complementation map of the *rudimentary* gene

complementation units were identified (figure B6.2). These four units were later shown to define discrete functional domains of this gene.[1] Region D of Rawls and Porter corresponds to region VII of Carlson; region C of Rawls and Porter corresponds to region VI of Carlson; region B of Rawls and Porter corresponds to region III of Carlson; and region A of Rawls and Porter corresponds to region I of Carlson. The genetic map and the complementation map are collinear. However, in Rawls and Porter's analysis, non-complementing alleles map to both the 5′ and 3′ ends of the gene.

Early findings that *r* mutants had a specific nutritional requirement for pyrimidines suggested that this locus might function in some component of pyrimidine biosynthesis pathway. Indeed, the fully non-complementing *r* mutants fail to produce three enzymes involved in this pathway: carbamyl phosphate synthase (CPSase), dihydroorotase (DHOase), and aspartate transcarbamylase (ATCase). Further biochemical studies revealed that all three of these enzymatic domains are produced as part of a trifunctional polypeptide

translated from a single mRNA. Measurement of these enzyme levels in the presence of the various complementing *r* mutants obtained by Rawls and Porter resulted in the following observations:

1. Mutants in complementation unit A resulted in a significant decrease in the activity of ATCase.
2. Mutants in complementation unit C resulted in a significant decrease in the activity of CPSase.
3. Mutants in complementation unit D resulted in a significant decrease in the activity of DHOase.
4. Non-complementing *r* mutants fail to produce all three enzymes involved in this pathway.

Thus the genetics alone predicted that there would be a trifunctional protein produced by the *r* message and that the order of domains would be DHOase → CPSase → ATCase. A study of putative regulatory mutations of the *rudimentary* gene confirmed the hypothesis that the three enzyme activities encoded by the locus (carbamyl phosphate synthase, aspartate transcarbamylase, and dihydro-orotase) are part of a trifunctional polypeptide, or that their genes are transcribed together as parts of a multicistronic transcript (Tsubota & Fristrom 1981). Indeed, the *r* locus has been cloned and multiple mutants have been sequenced (Segraves et al. 1984; Freund et al. 1986). The three enzymatic activities

---

[1] The rudimentary locus does indeed encode a fourth enzymatic activity, GATase, but as this activity was not examined in any of the earlier genetic studies, we have omitted it from consideration here.

Box 6.1   *(cont'd)*

are organized on the peptide chain in the following order: $NH_2 \rightarrow$ DHOase $\rightarrow$ CPSase $\rightarrow$ ATCase $\rightarrow$ COOH [Freund and Jarry (1987), as corrected in Lindsley and Zimm (1992)]. As predicted: complementation unit D corresponds to the DHO domain; unit C identifies the CPS domain; and unit A of Rawls and Porter identifies the ACT domain. The genetic analysis of *rudimentary* is the epitome of gene fine-structure analysis in Drosophila. Still, the amount of effort needed for this analysis can no longer be justified in view of modern molecular tools.

Meiotic recombination

Once you have your mutants in hand, you will want to map them. The most common method for doing so is to use meiotic recombination. For that reason, we present here an account of both the meiotic process itself and the mechanism and analysis of recombination.

7.1 An introduction to meiosis

Most people are introduced to meiosis by a slide similar to that shown in figure 7.1.[1] The slide is filled with obscure (and often wrong) images and hard-to-spell Latin words. This is unfortunate, because the basic events of meiosis are quite simple. First, homologous pairs of chromosomes must be identified and matched. This process, which occurs only in the first of the two meiotic divisions, is called **pairing**. The matched pairs must be physically interlocked by **recombination** (also called **exchange** or **crossingover**). Finally, at the first meiotic division, homologous chromosomes separate from each other and are partitioned into the two products of the first meiotic division. We describe this process simply as:

Match them   *and*   Lock them   →   Move them
  (Pairing)   (Recombination)   (Segregation)

The second meiotic division begins with half of the regular number of chromosomes. During the second meiotic division, sister chromatids of each chromosome separate into two daughter nuclei. Meiosis II may then be thought of as simply a haploid mitosis.

The most critical issues to understand about meiosis are the following eight assertions:

1 *Meiosis is the physical basis of Mendelian inheritance.* The rules of Mendelian inheritance are simply a restatement of the fact that all diploid organisms possess two copies of each chromosome. They transmit one, and only one, of those two chromosomes to their offspring as a consequence of anaphase I. Mendel's law of **independent assortment** says that for two genes **A** and **B** (that lie on different chromosomes), an individual of the genome **AaBb** will

---

[1] This slide is usually sufficient to scare most budding geneticists into other majors.

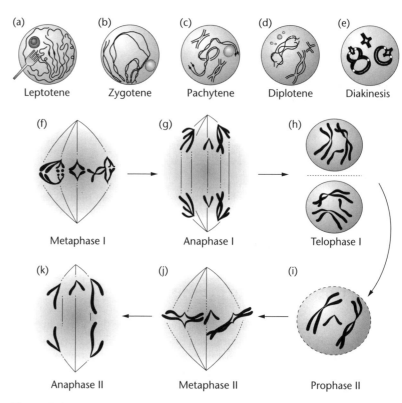

(a) Leptotene
(b) Zygotene
(c) Pachytene
(d) Diplotene
(e) Diakinesis

(f) Metaphase I
(g) Anaphase I
(h) Telophase I

(k) Anaphase II
(j) Metaphase II
(i) Prophase II

Figure 7.1   Meiosis made complex: ppt slide, identification of genetic material

form the gamete types **AB**, **Ab**, **aB**, and **ab** with equal frequency. This is truly solely because the chromosome pair bearing the **A** gene alleles orients independently of the homolog pair bearing the **B** gene; **AB** ↔ **ab** segregations are as likely as **Ab** ↔ **aB** segregations (figure 7.2).

2   *Linkage results from situations where the two gene pairs lie at different positions on homologous chromosomes.* In this case, recombinant gamete types (**aB** or **Ab**) can result only from recombination between the two homologous chromosomes.

3   *The function of recombination is to ensure homolog segregation at anaphase I.* The mechanism by which recombination events become chiasmata and ensure homologous segregation is fully detailed in Nicklas (1974) and summarized below. It may well be that recombination has long-term consequences for the fitness of newly arising mutants [cf. Barton and Charlesworth (1998)]. But within the context of understanding the meiotic process, exchange functions only to ensure homologous segregation.[2]

---

[2]   To say that the function of exchange is to increase or reduce genetic variation is a bit like suggesting that automobiles were created to produce smog – they do so, but only as a byproduct of their real function, moving people around.

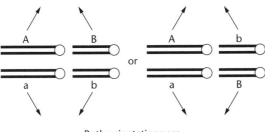

Both orientations are
equally likely

Figure 7.2 Independent assortment

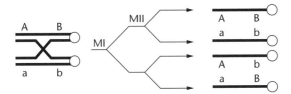

Figure 7.3 A bivalent with a single exchange. A single exchange involves only two non-sister chromatids. It produces two crossovers and two non-crossover chromatids. (Adapted from Botstein & Maurer 1982)

4   *The result of an exchange is the physical interchange of two non-sister chromatids (figure 7.3).* Each of the two resulting crossover chromatids is now connected to its sister on one homolog proximal to the crossover and to its sister on the other homolog distal to the exchange. This structure is referred to as a meiotic **bivalent**. Because sister chromatids are maintained proximally and distally to the exchange, the crossover event remains locked in position and serves to bind the two homologous chromosomes together (Buonomo et al. 2000; van Heemst & Heyting 2000; Lee & Orr-Weaver 2001). In this sense, we can say that exchange utilizes the sister chromatid adhesion that flanks the exchange event to create homolog adhesions. It is these adhesions that hold the bivalent together and ensure proper chromosome segregation at the first meiotic division.

5   *The failure of two homologs to undergo genetic exchange increases the likelihood that they will fail to segregate from each other properly at anaphase I (i.e. they will **nondisjoin**).* Organisms such as Drosophila (Hawley & Theurkauf 1993; Hawley et al. 1993), yeast (Dawson et al. 1986; Guacci & Kaback 1991; Molnar et al. 2001), and *C. elegans* (Albertson et al. 1997) certainly have back-up systems to ensure the segregation of those homolog pairs that might fail to recombine. It is likely, but not yet proven, that such systems exist in mice and humans as well. However, in all those organisms, the back-up systems have a lower degree of fidelity than does exchange-based segregation. Moreover, back-up systems clearly do not exist in at least some other organisms, such as higher plants. In those organisms homologous chromosomes that do not undergo exchange dissociate from each prior to the first meiotic division and segregate at random, with respect to each other, at that division. Indeed, in most organisms so far studied the majority of spontaneous **nondisjunction** (failed segregation) results from the misbehavior at

Chapter 7

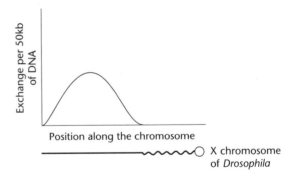

Figure 7.4 Exchange per unit length

meiosis I of pairs of homologous chromosomes that either failed to undergo exchange or in which the exchange(s) that did occur were positioned close to the telomeres [Koehler et al. (1996); Lamb et al. (1996); but see below]. Nondisjunction can also occur at the second meiotic division, but this occurs far less frequently in most organisms.

6 *The number and distribution of exchange events is not random, but rather, tightly regulated.* All chromosome pairs will have an average of more than one crossover per bivalent, regardless of their size. Exchange does not occur in the centric heterochromatin, and is reduced near the telomeres and in the vicinity of the centric heterochromatin. A plot of exchange distributions per DNA length is presented in figure 7.4. Moreover, the occurrence of a one-exchange event within a given chromosomal region decreases the probability that a second exchange will occur within that region. This phenomenon is referred to as **crossover interference**.

7 *The initiating event for recombination is a **double-strand break** (DSB) created by a Spo-11 type enzyme.* This feature of meiotic recombination appears to be universally conserved (Lichten 2001). DSB formation is required for synapsis in yeast [cf. Roeder (1997)], Arabidopsis (Grelon et al. 2001), and in the mouse (Baudat et al. 2000; Romanienko & Camerini-Otero 2000; Mahadevaiah et al. 2001). However, the dependence of synapsis on DSB formation may be plastic in terms of evolution in the sense that DSB formation is not required for synapsis in either Drosophila (McKim & Hayashi-Hagihara 1998; McKim et al. 1998) or *C. elegans* (Dernburg et al. 1998). A further discussion of the relationship between the initiation of recombination and synaptonemal complex (SC) formation is presented in box 7.1.

8 *The mechanisms by which chromosomes pair and synapse remain obscure.* We can say only that once synapsed, meiotic bivalents are connected together by a structure known as the synaptonemal complex that runs along their length. Evidence that pairing may, or may not, be initiated at specific sites in various organisms is summarized in box 7.2.

The purpose of this chapter is to review the evidence that underlies these assertions in sufficient detail to make it clear to the reader how the physical events of meiosis allow the use of meiosis as a genetic tool (cf. mapping and segregation analysis). It is not our intent to review the process of meiosis in general.

Box 7.1   The molecular biology of synapsis

When geneticists and cytologists discuss the processes that bring homologous chromosomes together during meiotic prophase, we often use the words alignment, pairing, and synapsis. Unfortunately, these words do not always mean the same thing, but it still seems worthwhile to attempt to provide some sort of definitions here. By **alignment**, most investigators refer to the process of bringing homologous chromosomes into rough apposition along their length. The term **pairing** refers to an intimate physical association of the homologs. The term **synapsis** refers to the stage in which the paired homologs are connected along their length by a railroad track-like structure referred to as the **synaptonemal complex**. This is a tripartite structure consisting of two lateral elements that flank the chromosomes, and a central element. Lateral elements are derived from the axial cores of meiotic chromosomes. They are initially assembled between the two sister chromatids of the leptotene chromosome and then move to lie to one side of both sister chromatids [cf. Wettstein (1984)]. Indeed, one can think of the lateral elements of the SC as a permutation of the cohesin-type complexes that mediate sister chromatid cohesion. An excellent review of the structure of the SC may be found in Zickler and Kleckner (1999).

In the 1990s, *S. cerevisiae* geneticists reported the surprising observation that mutational ablation of the SC had surprisingly mild consequences for exchange and segregation (Rockmill & Roeder 1990). Indeed in *S. cerevisiae*, the initiation of both recombination and chromosome pairing occurs in early leptotene, concomitant with or before synapsis (Roeder 1997). Yeast mutants that completely suppress recombination also block the formation of a synaptonemal complex. In contrast, other yeast mutants can block synapsis while allowing near normal levels of gene conversion and only two- to threefold reductions in meiotic recombination. As noted by Roeder (1997), in yeast "synapsis is not required for recombination; instead steps in the recombination pathway appear to be required for synapsis. Mutants that do not sustain DSBs fail to make SC." Similar conclusions have been reached in Arabidopsis (Grelon et al. 2001) and in the mouse (Baudat et al. 2000; Romanienko & Camerini-Otero 2000; Mahadevaiah et al. 2001). Moreover, recombination is initiated quite normally in *S. pombe* (which does not assemble an SC during meiosis) and is apparently fully sufficient to ensure segregation.

However, it was suggested by McKim et al. (1998) that exchange was not always necessary to support synapsis (they demonstrated that exchange is not required for synapsis in Drosophila). The *mei-W68* and *mei-P22* mutations in Drosophila both reduce the frequency of meiotic recombination by more than 1,000-fold. Despite the complete absence of recombination in the *mei-W68* and *mei-P22* oocytes, the assembly of the synaptonemal complex appears to proceed normally (McKim et al. 1998). This observation became all the more striking in light of the subsequent finding that the *mei-W68* gene encodes the Drosophila homolog of Spo11p, the yeast protein required for double-strand break formation (McKim & Hayashi-Hagihara 1998). Similar observations were made in *C. elegans* by Dernburg et al. (1998).

How then are we to deal with this rather disturbing inconsistency of functional relationships? Zickler and Kleckner have suggested that all of these observations are compatible with a view in which "the commitment of a recombinational interaction to a crossover fate is tightly coupled with the formation of an underlying SC patch which would then nucleate formation of SC." They explain the Drosophila and *C. elegans* exceptions by suggesting that strong secondary mechanisms may exist to facilitate SC formation in organisms like *C. elegans* and Drosophila. We note in box 7.2 that *cis*-acting chromosomal sites, possessing the genetic properties expected of pairing initiating or stabilization sites, have been found in both Drosophila and *C. elegans*. Perhaps these sites can facilitate SC formation in flies and worms, even in the absence of exchange initiation. As noted by Zickler and Kleckner (1999), the meiotic programs of various organisms may "differ only in respect to the potency of secondary SC nucleation mechanisms, which can promote SC formation in

Box 7.1 (cont'd)

a normal time frame even when the normal nucleation mechanisms are missing."

Our colleague, Dr. James Haber, has suggested that the existence of such strong secondary SC initiating mechanisms in Drosophila and *C. elegans*, and their apparent absence in yeast, might explain the discrepancies in the effects of heterozygosity on exchange initiation in these organisms. Although dispersed homologous DNA sequences undergo meiotic recombination and gene conversion at high frequencies in yeast (Jinks-Robertson & Petes 1985, 1986; Lichten et al. 1987), the same is not true in Drosophila (Grell 1964; Hawley 1980; Hipeau-Jacquotte et al. 1989). Even large

chromosomal regions fail to recombine and/or synapse with their homologous intervals when they are translocated to other positions in the genomes of Drosophila (Craymer 1981, 1984) and *C. elegans* (McKim et al. 1988a,b, 1993). However, rearrangement heterozygotes which block both synapsis and recombination in most higher organisms have little effect on recombination in yeast (Sherman & Helms 1978). Haber suggests that the rapid formation of SC in flies and worms, by an exchange-independent mechanism, may preclude the ability of homologous sequences on non-homologous chromosomes from being able to interact with their partners.

Box 7.2   Do specific chromosomal sites mediate pairing?

As reviewed by Loidl (1990), a number of workers have proposed that various chromosomal organelles may play important roles either in initiating SC formation or in maintaining synapsis. Two types of specific pairing elements or putative pairing sites, one in *C. elegans* and one in Drosophila, have also been identified by genetic studies aimed at identifying sites required for normal levels of recombination. The studies presented below suggest that both chromosomal elements and these putative pairing sites may well define elements required to establish or maintain synapsis in these organisms.

The role(s) of telomeres in early pairing

Cytogenetic studies of the first meiotic prophase of many organisms have demonstrated the clustering of telomeres (usually in the vicinity of the centrosome or the spindle pole body) during the leptotene → zygotene transition [reviewed by Chikashige et al. (1994, 1997); Dernburg et al. (1995); Scherthan (1997); Rockmill and Roeder (1998); Zickler and Kleckner (1998); Trelles-Sticken et al. (1999)]. Because this configuration, known as a chromosomal bouquet, precedes the initiation of synapsis, the possibility has been raised that this early localization of the telomeres to a small region of the nuclear envelope facilitates the alignment of

homologous chromosomes. In addition, *S. pombe* mutants that disrupt telomere clustering reduce the frequency of recombination (Cooper et al. 1998; Nimmo et al. 1998), suggesting that telomere clustering acts to facilitate homolog alignment.

Although telomere clustering may provide a useful step in early chromosome alignments, the clustering of a large number of telomeres might also present problems in the creation of individualized bivalents. This possibility is suggested by the studies of Chua and Roeder (1997) and Conrad et al. (1997) on the NDJ1 gene in *S. cerevisiae*. NDJ1 encodes a protein that is required for the completion of homologous synapsis, and which accumulates at the telomeres of chromosomes during meiotic prophase. Loss of the NDJ1 protein delays the formation of the axial elements of the synaptonemal complex and results in high levels of failed meiotic chromosome segregation. Moreover, there is no effect of the absence of the NDJ1 protein on the segregation of telomere-less ring chromosomes, arguing that the NDJ1 protein is not required for meiotic chromosome separation *per se*, but rather that the NDJ1 protein is essential to separate segregational partners that have telomeres. The mechanism by which the NDJ1 protein facilitates synapsis remains unclear, but one could imagine that the normal clustering of telomeres into a single bouquet might create three-dimensional chromosome arrangements, such as interlocked bivalents, that would impede proper synapsis.

Role(s) of the centric heterochromatin in chromosome pairing

The importance of heterochromatic pairings in mediating pairing and segregation was first suggested by studies of the effects of homologous duplications on the segregation of the achiasmate chromosomes in Drosophila oocytes (Hawley et al. 1992). This hypothesis was verified by the work of Karpen et al. (1996), who showed that the frequency with which two deletion derivatives of a mini-chromosome segregate from each other during female meiosis is directly proportional to the amount of centric heterochromatic homology shared by the pairing partners. Normal segregation was shown to require that the two mini-chromosomes share 800 kb of overlap in the centric heterochromatin; nearly random disjunction was observed when the two mini-chromosomes shared only 300 kb of heterochromatic homology. Dernburg et al. (1996) used three-dimensional fluorescence *in situ* hybridization to demonstrate that although euchromatic pairings not locked in by chiasmata quickly dissolve following pachytene, heterochromatic pairings are still preserved.

Specific pairing sites in *C. elegans*

Two types of genetic studies of chromosome rearrangements in *C. elegans* have suggested that a single region (known as the homolog recognition region, or HRR) exists at the end of each chromosome, and that this region may play a primary role in the pairing of homologous chromosomes (Rosenbluth & Baillie 1981; McKim et al. 1988a,b). The central observation is that when a chromosome is split by translocation or large deletion, only the pieces that maintain the HRR are capable of pairing in a sufficiently stable manner to allow them to recombine with their homologs. For example, only duplications that contain an HRR are capable of recombining with their intact homolog (Herman & Kari 1989; McKim et al. 1992). Similarly, in worms heterozygous for a reciprocal translocation, crossing over is limited to regions of each chromosome that are physically contiguous with the HRR elements (McKim et al. 1988a,b, 1993). Based on both the genetic and cytological data, HRRs have been proposed by several workers to function as major synapsis sites [cf. Villeneuve (1994)].

Specific euchromatic pairing sites in Drosophila

Studies in Drosophila have also pointed to a role for internal sites or elements that facilitate stable pairing and synapsis. In her Master's Thesis in 1956, Iris Sandler examined X chromosomal exchange in females heterozygous for a normal sequence X chromosome and for one of two different translocations between the X and 4th chromosomes. She observed that the X chromosome could be divided into large intervals, such that heterozygosity for a breakpoint within one interval strongly suppressed recombination within that interval but did not suppress exchange in adjacent intervals. She proposed that so-called "pairing sites" bound these intervals and that both homologous chromosomes must be continuous between two sets of paired sites for normal levels of exchange to occur within the interval bounded by those sites.

Hawley (1980) refined these observations by analyzing 20 more rearrangements, and precisely mapping the proposed sites. Hawley also tested a series of duplications for these sites and found several genetic properties that might be expected for such pairing sites. As shown in figure B7.1, such duplications suppress exchange only in intervals for which they carry a boundary or pairing site. Indeed, free duplications for a putative "pairing" site can suppress exchange throughout any interval bounded by that site, but not beyond the next pairing site(s) (Hawley 1980). In the absence of the completion of cytological studies of pairing in translocation heterozygotes in Drosophila, it is not possible to discern whether these sites function at the level of pairing or synapsis. However, given the *C. elegans* studies described above, we currently favor the possibility that these sites may also play a role in the maintenance of synapsis.

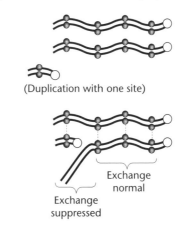

(Duplication with one site)

Exchange
normal

Exchange
suppressed

Figure B7.1 Duplications

The interested reader is referred to several excellent reviews of meiosis in a variety of organisms for that purpose (Hawley et al. 1993; Roeder 1997; Zickler & Kleckner 1998, 1999; Davis & Smith 2001; Hassold & Hunt 2001). However, for purposes of discussion we do need to provide a brief summary of the meiotic process.

7.1.1 A cytological description of meiosis

As shown in figure 7.5, the two divisions of a generic meiosis may be divided into five steps. The term **meiotic prophase** refers to the period after the last cycle of DNA replication. It is during this interval that homologous chromosomes pair and recombine. (Pairing and recombination do not reoccur during any part of meiosis II.) The breakdown of the nuclear envelope signals the end of meiotic prophase. The term prometaphase is used to describe the period during which the bivalents attach to (animal males) or create (animal females) the meiotic spindle and congress to the center of that spindle. **Metaphase I** is defined as the period before the first division during which the **bivalents** are lined up on the middle of the meiotic spindle, a position referred to as the **metaphase plate**. The chromosomes are primarily (but not exclusively) attached to the spindle at the **centromere**. The centromere of one homolog is attached to one pole, and the centromere of its partner is attached to the other pole (see figure 7.3). These bivalents are physically held together by **chiasmata**, the physical manifestation of meiotic recombination events. In most meiotic systems, meiosis will not continue until all of the homolog pairs are properly oriented on the **metaphase plate**.

Once the homologs are properly aligned, sister chromatid cohesion is released along the arms of the meiotic chromosomes, but *not* in the region surrounding their centromeres. This event frees the two chromosomes in each bivalent from their chiasmate attachments, and thus allows each chromosome to proceed towards the closest spindle pole. This event is referred to as the **metaphase–anaphase transition**, and initiates **anaphase I**. Because the release of sister chromatid cohesion at the metaphase–anaphase transition of meiosis I occurs *only* along the arms of the bivalent, and not at the centromeres, each homolog is still comprised of two sister chromatids held together by cohesion at the centromeres.

Some organisms culminate the first meiotic division with a true **telophase** (a time in which nuclei reform), while others simply proceed directly into meiosis II. In those organisms that do reform nuclei at the end of meiosis I there may also be a brief **prophase II**. DNA replication does not occur during prophase II; each chromosome still consists of the two sister chromatids. There are no opportunities for pairing or recombination at this stage, due to the prior separation of homologs at anaphase I.

Following the completion of the first meiotic division, the chromosomes of each meiosis I product align themselves on a new pair of spindles with their sister chromatids oriented towards opposite poles. This stage is referred to as **metaphase II**. **Anaphase II** is signaled by the separation of sister centromeres,

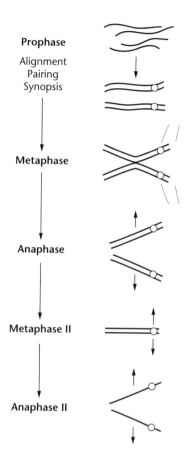

Prophase

Alignment
Pairing
Synopsis

Metaphase

Anaphase

Metaphase II

Anaphase II

Figure 7.5 Schematic of the meiotic cell cycle

and the movement of sister chromatids to opposite poles. At **telophase II**, the sisters have reached opposite poles and nuclei begin to reform. Thus, at the end of the second meiotic division, there are four daughter nuclei, each with a single copy of each chromosome.

We feel obligated to note that the thumbnail description of meiotic cytology presented above belies the enormous diversity that has been documented with respect to the meiotic process. For example, female meiosis in many (but not all) animals is acentriolar (at least for meiosis I); the chromosomes themselves organize an aster-less meiotic spindle (Theurkauf & Hawley 1992). Female meiosis in Drosophila also has an unusual end to prophase, in which the chromosomes condense into a dense mass, where they remain until the beginning of prometaphase. Finally, in yeast, the entire meiotic and mitotic process takes places without nuclear envelope breakdown. But more critically, this variation in common experimental systems is quite small given the large number of unusual and often bizarre meiotic systems documented by insect cytogeneticists in the last century [cf. Wolf (1994)].

7.1.2   A more detailed description of meiotic prophase

Because pairing and recombination occur during the first meiotic prophase, much attention has been focused on this stage of the process. [For an excellent review of the events of meiotic prophase the reader is referred to Zickler and Kleckner (1998).] The prophase of the first meiotic division is subdivided into five stages: **leptotene, zygotene, pachytene, diplotene**, and **diakinesis**. The term **leptotene** describes the initial phase of chromosome individualization, during which initial homolog alignments are made, while by **zygotene**, homologous chromosomes are associated at various points along their length. There are three critical events that occur during leptotene and zygotene. These are: (i) homolog recognition and alignment; (ii) the formation of a track-like structure called the synaptonemal complex that connects the paired chromosomes along its entire length; and (iii) the initiation of meiotic recombination.

The mature SC is a tripartite structure consisting of two lateral elements, which flank the chromatin, and a single central element. Lateral elements are derived from the axial cores of meiotic chromosomes. The function of the SC remains obscure (see boxes 7.2 and 7.3). Although SC formation may be essential to initiate recombination in organisms such as Drosophila (Walker & Hawley 2000), meiotic recombination occurs quite happily in *S. pombe* in which no SC is formed (Davis & Smith 2001). In organisms that do possess SC, sites of recombination are marked along the meiotic chromosomes during pachytene by structures sitting on top of the SC. These spherical structures are known as late recombination nodules. Adelaide Carpenter discovered late recombination nodules in the 1970s (Carpenter 1975, 1979a,b, 1981, 1984). In doing so, she demonstrated that their number and distribution parallels that of exchange events. Evidence that late nodules do indeed mature into chiasmata is reviewed by Wettstein (1984), Carpenter (1987, 1988), and Hawley (1988).

The beginning of **pachytene** is signaled by the completion of a continuous SC running the full length of each bivalent and its end is signaled by the dissolution of the SC. During **diplotene**, the attractive forces that mediated homologous pairing disappear, and the homologs begin to repel each other. At this stage in most organisms, homologs are held together only by their chiasmata. Indeed, in organisms without a back-up system for recombination, those rare chromosome pairs that have failed to undergo recombination will fly apart prematurely from each other at this stage. The final stage in meiotic prophase is **diakinesis**, during which the homologs shorten and condense in preparation for nuclear division.

7.2   Crossingover and chiasmata: recombination involves the physical interchange of genetic material and ensures homolog separation

Chiasmata can be visualized at diplotene–diakinesis as sites on the bivalent in which two non-sister chromatids appear to cross over from one homolog to the other. At the beginning of the twentieth century there were two views of the

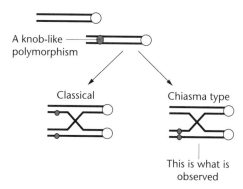

Figure 7.6 Classical versus chiasmatype hypothesis

origin of chiasmata. According to the chiasmatype hypothesis of Janssens (1909), chiasmata are the cytological manifestation of genetic exchanges. According to a competing hypothesis, the "classical hypothesis" chiasmata were simply sites at which chromosomes have traded sister chromatid regions without breakage and reunion. In these hypothetical structures, the chromatid swapping events can be resolved by exchanges but are not the consequences of them. These two models are portrayed in figure 7.6. There are numerous proofs that the chiasmatype hypothesis is correct [for a review, see Whitehouse (1982)]. However, the most compelling of these involved the cytological analysis of bivalents that were in some way heteromorphic. For example, the two homologs might differ by a heterochromatic knob or a block of heterochromatin at their tips. As shown in figure 7.7, the two models make very different predictions for the structure of such chiasmate bivalents, and it is the structure consistent with the chiasmatype hypothesis that is always observed.

The chiasmatype model predicts that exchanges will result in the physical interchange of homologous material. Thus one predicts that the two chromosomes produced by the type of exchange diagrammed in figure 7.7 will be heteromorphic after the bivalent has separated at anaphase I. Thus each chromosome that experiences a crossover proximal to the heteromorphic marker will possess a chromatid of each type (e.g. knobbed or knobless). Exactly such heteromorphic dyads have been observed at anaphase by a number of workers [cf. Maguire et al. (1991)]. Finally, the inheritance of interchanges of distal knobs and heterochromatic blocks following exchange was documented by Creighton and McClintock (1931) in maize and by Stern (1936) in Drosophila. Stern's experiment is summarized in figure 7.8. These experiments demonstrated unambiguously that chiasmata were the result of exchange. But it was the work of Bruce Nicklas (1974) that demonstrated exchanges serving the vital function of ensuring homolog separation at the first meiotic division. These experiments are described in detail in chapter 8 (section 8.4.1).

Figure 7.7 A heteromorphic bivalent makes two heteromorphic dyads at anaphase I

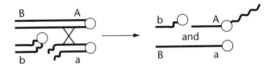

Figure 7.8   Crossingover involves the physical exchange of genetic material between the two homologs, not only the genetic markers (A and B) but the aberrations as well

7.3   The classical analysis of recombination

Larry Sandler began his Chromosome Mechanics class at the University of Washington by handing out a document he referred to as the *Ten Commandments of Crossing-Over*. A modified and annotated version of this list, presented below, provides an excellent summary of the formal "rules" of meiotic exchange.

1   *In each chromosome the genes are arranged in a linear series, and corresponding groups of genes are exchanged in crossingover.* This is self-evident to the modern biologist. To the early geneticist, however, this conclusion derived from the fact that it was possible to map three genes in a linear order (three-factor mapping).

2   *Exchanges are complementary and involve the physical exchange of material.* The evidence for this assertion is presented above.

3   *Exchanges occur when each chromosome consists of exactly two chromatids.* This is to say that recombination occurs after DNA replication. There are proofs of this assertion in many organisms, most notably in Neurospora (Perkins 1962) and in Drosophila (Anderson 1925; Beadle & Emerson 1935). A comprehensive discussion of this problem is found in Zhao et al. (1995). The evidence in Drosophila is based on a truly elegant analysis of recombination within compound X chromosomes, as presented in box 7.3.

4   *Each exchange event involves only two chromatids, one from each chromosome.* This is a critical rule because it reminds us that a bivalent with a single exchange will produce an equal number of crossover (two) and non-crossover (two) chromatids (see figure 7.3).

5   *A given bivalent may undergo more than one exchange event, but for each such event crossingover is limited to two chromatids, again one from each chromosome.* This statement means that a single paired bivalent may be involved in one or more than one exchange event. Thus, single, double, and triple (etc.) crossover products are possible. However, any given exchange can involve only two chromatids, one from each chromosome.

6   *In bivalents with two or more exchanges, the choice of the two chromatids that crossover in one exchange does not affect the choice of which two chromatids participate in the other exchange(s).* In other words, chromatid choice is random (except for sister exclusion) for each of the exchanges. This rule is often referred to as the "*no chromatid interference*" rule. The easiest way to understand this rule is to look at the four classes of double crossover bivalents shown in figure 7.9. In the uppermost bivalent, referred to as a *two-strand double*, both crossovers involve the same two chromatids. In the two middle

Box 7.3   Crossingover in compound X chromosomes

L. V. Morgan recovered the first attached X chromosome on February 12, 1921. It is also referred to as a compound X chromosome. This chromosome, perhaps more properly designated an isochromosome, consisted of two full-length X chromosomes whose left arms appear to be fused at a medial centromere. (There are no essential genes on the very small right arm of the fly X chromosome. Thus a female carrying only an attached X chromosome is fully viable and fertile.) Presumably this chromosome arose by a translocation-like event with one break on the right arm of one chromosome and the other very proximal on the left arm of the other X chromosome. The rejoining then attached virtually the entire left arm of the X chromosome to the centromere of the other. We refer to this type of attached X as a *C(1)RM* chromosome (for *compound-one-reversed metacentric*). Other types of compound X chromosomes exist, and their recombinational properties have also been studied in detail (Lindsley & Sandler 1963), as have those of compound chromosomes carrying two copies of the four autosomal arms in Drosophila (Holm 1976).

However, for our purposes here we need only consider the original type of compound X chromosome. This chromosome segregates as a univalent at meiosis in females with no other sex chromosomes, and thus it segregates to a single pole at anaphase. At meiosis II, the two sister chromatids comprising the compound chromosome divide normally, transmitting a full copy of the compound X (as a single chromatid) to each daughter cell. Thus, the compound is transmitted to only half of the ova; the remaining ova receive no X chromosomal material.

We are concerned here with the case in which the two arms used to construct the compound X chromosome are heterozygous for multiple genetic markers (figure B7.2). Note that a single crossover between the arms of the compound X chromosome can result in the formation of a crossover product that is homozygous for all markers distal to that crossover event (Anderson 1925; Beadle & Emerson 1935). This fact alone demonstrates that exchange must occur at the four-strand stage. Exchange at the two-strand stage simply cannot result in homozygosity.

One can thus easily detect those cases where a given mutant marker has become homozygous. But how does one distinguish the homozygous of the wildtype marker from a retention of heterozygosity. The ability to passage the attached X chromosome in question through meiosis in several subsequent generations of females allows us to fully determine the genotype of any such chromosome. If that marker is still heterozygous, it will eventually be revealed as a homozygote by a marker proximal crossover event. This process of "genotyping" an attached X after one passage through meiosis allows us to determine the type of tetrad from which it arose. This then allows us to perform a reasonably complete tetrad analysis.

The frequency with which any given recessive marker can be made homozygous will increase with its distance from the centromere. (This observation allowed the first accurate mapping of the X centromere in Drosophila.) If there is no sister

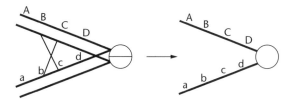

Figure B7.2   Exchange between the arms of an attached X chromosome can result in homozygosity for distal markers

Box 7.3   (*cont'd*)

chromatid exchange, then the maximum frequency of homozygosis ought to be 25%. However, if sister chromatid exchange were to occur with the same frequency as does inter-homolog exchange, then the maximum frequency of homozygosis would be 16.6% (figure B7.3). As can be seen from the figure, the frequency of homozygosis for distal markers clearly and repeatedly exceeds 16.6%, and approaches the maximum value of 25% expected if no sister chromatid exchange is allowed. Then, as expected from the accumulation of double and higher levels of exchange, the frequency of homozygosis of recessive markers will decline towards a final value of 16.6%.

The logic here is identical to that presented in the text for the relationship of second division

segregation frequencies to map length. Over long distances the double (and higher level) exchanges will effectively randomize the order of the four alleles of each marker within the attached X. So if you had four alleles of a very distal gene, **AAaa**, in your hand, now assign the **a** allele to the first of the four chromatids. The odds that the other arm of the same chromatid attached X would pick up the **a** allele are only a third. For this reason the maximum possible frequency of marker homozygosis (both **AA** and **aa**) is going to be 33.3%, and recessive homozygotes will only account for half of these.

Again the critical point is that the medial values in this curve did exceed 16.6%. This would not be expected if sister chromatid exchanges competed with inter-homolog exchanges.

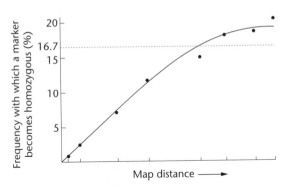

Figure B7.3   The relation of percentage homozygosis to map distance

bivalents, referred to as a *three-strand double*, one chromatid is involved in both crossover events. In the lower bivalent, a *four-strand double*, different pairs of chromatids are involved in each of the two crossover events. One might imagine that if a given chromatid was involved in the first exchange event, it might be more or less likely to participate in the second. Indeed, this is not at all the case. There is no chromatid interference or preference in the second event, as shown by the fact that the ratio of two-, three-, and four-strand doubles is very close to the predicted ratio of 1:2:1. As described below, this has been clearly and directly demonstrated in Neurospora (Perkins 1955, 1962) and flies (Anderson 1925; Beadle & Emerson 1935). See box 7.5.

(a) Two-strand doubles

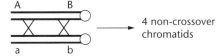

→ 4 non-crossover chromatids

(b) Three-strand doubles

→ 2 crossover chromatids
2 non-crossover chromatids

(c) Four-strand doubles

→ 4 crossover chromatids

Figure 7.9 The three classes of double crossovers

7 *Meiotic recombination does not occur between sister chromatids; or if it does, it does not interfere with recombination between homologs.* Because exchange between sister chromatids generally has no genetic consequences, it has been difficult to determine whether or not it occurs meiotically, and if so how often. Studies designed to examine sister chromatid recombination during meiosis are presented in box 7.4.

8 *There is crossover (or regional) interference.* In many meiotic systems the occurrence of an exchange in one interval interferes with the occurrences of other exchanges in neighboring intervals. In some organisms, and on some chromosomes, this interference can be strong enough to limit exchange to approximately one crossover per bivalent. Interference usually decreases with distance and finally vanishes. The mechanism of regional or "crossover" interference remains a mystery, although several good models have recently been proposed (Foss et al. 1993; Foss & Stahl 1995).

9 *The frequency of recombination is not simply proportional to the unit length of DNA, but rather is under tight genetic control from both cis and trans-acting elements.* As shown in figure 7.4, the amount of exchange per chromosome arm is not uniform. Exchange levels are reduced in the proximal and most distal euchromatin, and absent in heterochromatic intervals. However, even within small euchromatin intervals, careful studies of sites of exchange initiation reveal obvious hot-spots and cold areas for DSB formation (Wu & Lichten 1994, 1995; Kirkpatrick et al. 1999).

10 *Exchanges are generally sufficient to ensure the segregation of two chromosomes, even if their centromeres are non-homologous.* Although very distal exchanges do not always ensure segregation (Carpenter 1973; Rasooly et al. 1991; Koehler et al. 1996; Lamb et al. 1996; Ross et al. 1996a), in the vast majority of cases one meiotic exchange will ensure that two chromosomes segregate from each other. This is true even in those cases, such as in a translocation heterozygote, where the homologous intervals that participated in the exchange are attached to non-homologous centromeres.

Box 7.4　Does any sister chromatid exchange occur during meiosis?

Throughout this chapter we have used multiple lines of evidence to demonstrate that either sister chromatid exchange (SCE) does not occur, or if it does, it does not compete with inter-homolog exchanges. Still this does not answer the question: does sister chromatid exchange occur at a measurable frequency during meiosis? During the last two decades, several workers have approached this question in Drosophila and yeast by genetic means. More recently, Kleckner and her collaborators have addressed the question by more direct molecular approaches. Both approaches suggest that while sister chromatid recombination events do indeed occur, they are substantially less common than are inter-homolog recombination events.

Genetic studies in yeast

As shown in figure B7.4, SCE change that occurs within a ring chromosome is detectable because it produces a dicentric chromosome that is virtually always lost at the second meiotic (or first mitotic division). (Exchange between a ring and a normal sequence linear homolog can also create dicentric products. But these two classes of events are easily distinguishable in an organism like yeast, where one can recover all four products of meiosis. SCEs involving the ring will produce two viable spores, both carrying the linear chromosome. Exchange between the ring and the rod result in one viable rod- and one viable ring-bearing spore.) Based on such studies, Haber et al. (1984) estimated that the ring chromosome used in their studies underwent sister chromatid exchange in approximately 15% of the cell. This estimate of the frequency of

SCE in a ring was confirmed by Game et al. (1989), who used pulsed field DNA gels to detect the dicentric chromosomes produced by sister chromatid exchanges involving a ring chromosome in yeast. While this frequency of SCEs may seem high, the per meiosis frequency of SCE for this chromosome (0.15) is substantially lower than the per meiosis frequency of inter-homolog exchanges (1.7).

Genetic studies in Drosophila

Rita Khodosh (then an undergraduate in our laboratory) performed a similar ring-loss experiment in Drosophila, whose results were reported in McKim et al. (1998). In this experiment ring loss was measured by crossing females carrying a ring X chromosome and a balancer X chromosome [R(1)w^{vc}/FM7] to appropriate tester males. [R(1)w^{vc} is a ring X chromosome and FM7 is a multiply inverted X chromosome balancer chromosome that strongly suppresses inter-homolog exchange.] Khodosh assayed ring loss (presumably dicentric formation) by comparing the frequency of R(1)w^{vc}-bearing female progeny with that of the corresponding FM7-bearing sisters. In the case of otherwise genetically normal females, the Ring/Rod (FM7) ratio was 0.854 (N = 2,731). However, when Khodosh did the same experiment in females that lacked the fly equivalent of the SPO11 protein, the Ring/Rod ratio was increased to 1.079 (N = 659). Thus blocking the initiation of recombination seems to increase the frequency of ring recovery by approximately 15%. A very similar set of results was obtained by Jeff Hall using a second null

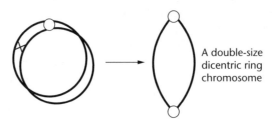

A double-size
dicentric ring
chromosome

Figure B7.4　SCE in a ring chromosome produces a dicentric ring recombination mutant *c(3)G¹⁷*. Using a different ring X chromosome [*R(1)2*], Hall observed a Ring/ Rod ratio of 0.755 (*N* = 7,552) in controls and an elevated ratio of 0.894 (*N* = 5,355) in *c(3)G* females. One reasonable explanation for these observations is that the reduced ring recovery observed in the two control experiments reflects the background frequency of meiotic sister chromatid exchange, and that *mei-P22* and *c(3)G* actually inhibit these sister chromatid exchange events, as they do inter-homolog events, and in doing so increase the transmissibility of the ring X chromosome.

A direct molecular assessment in yeast

We have already noted that Schwacha and Kleckner (1995) have succeeded in recovering double Holliday recombination intermediates, known as joint molecule recombination intermediates (JMs), produced during meiosis in yeast. Subsequently, Schwacha and Kleckner (1997) also examined the relative frequency with which such molecules involved homologs versus sister chromatids. Their data suggested that inter-homolog recombination events outnumber sister chromatid events by 2.4 to 1. Of perhaps equal interest is the observation that several yeast recombination mutants impaired this preference for inter-homolog events. These authors conclude, "most meiotic recombination occurs via an interhomolog-only pathway along which interhomolog bias is established early, prior to or during DSB formation, and then enforced just at the time of DSB formation. A parallel, less differentiated, pathway yields inter-sister events."

7.4   Measuring the frequency of recombination

There are multiple ways to determine the frequency of meiotic recombination. Perhaps the most direct approach is to simply count chiasmata [cf. Lawrie et al. (1995); Barlow and Hulten (1998)]. Unfortunately such estimates are not possible in all organisms, and they do not usually permit the mapping of specific markers along the length of chromosomes. Such genetic, or recombinational, mapping requires the recovery and assessment of crossover chromatids and the calculation of recombination distances (more commonly referred to as map lengths). In this section we discuss the tools required for using the frequency of crossover chromatids to do exactly that.[3]

7.4.1   *The curious relationship between the frequency of recombination and chiasma frequency (and why it matters)*

Imagine an organism in which the two copies of chromosome 1 underwent one, and precisely one, exchange per meiosis. In this meiosis the frequency with which the chromosome 1 bivalent would be chiasmate would also be 1.0. However, because that exchange involves only two of the four chromatids that comprise that bivalent, the resulting product of the meiosis will be two crossover

---

[3]   A brief note on terminology: Unfortunately, most geneticists now use the terms recombination, exchange, and crossing over interchangeably (no pun intended) to describe the process of meiotic recombination. Because we are as guilty of this as anyone, we cannot and will not try to change that practice here. But we will reserve the term **crossover chromatid** to mean *only* a chromatid that carries the non-parental combination of flanking markers.

and two non-crossover chromatids. Thus, despite a chiasma frequency of 1.0, the crossover frequency will be only 0.5. To express this as a formalism:

Chiasma frequency = Crossover frequency × 2

This matters for a number of reasons, including being able to estimate one parameter from the other. But most importantly, it matters because it reminds us that a non-crossover chromatid need not have come from an achiasmate bivalent. Indeed, a look at figure 7.9 will reveal that non-crossover chromatids can be obtained from two of the three classes of double exchange bivalents. Thus, simply knowing the number of chromatids with zero, one, two, or more crossovers does not by itself provide us with a picture of the exchange distribution at the meiosis that produced those chromosomes.

7.4.2   *Map lengths and recombination frequency*

If asked how to determine the map length between two markers on a given chromosome, most undergraduates would simply tell you how to measure the frequency of crossingover between those two markers. That is to say, they would answer "simply divide the total number of crossovers by the total number of progeny." The fraction of crossover is then multiplied by 100 to convert the data into recombination units, known as **map units** or **centimorgans** (cM). Thus, at least over short intervals, one map unit or 1 cM is equivalent to a 1% recombination. Obviously the maximum recombination distance, as assessed by using a single pair of markers, is 50 cM (at this point the markers display independent assortment). Maps with total lengths much greater than 50 map units can be constructed by doing many pairwise crosses for genes along a given chromosome arm and summing the map distances for all subintervals. In many organisms, such as Drosophila, where the average number of crossover events is low (i.e. those in which crossover interference is high), or the intervals being assessed are short enough that multiple exchanges are likely to be rare, this is a perfectly acceptable estimation of map length.[4]

If, however, one is performing a two-point (i.e. just two genes are being studied) measurement over large intervals, or working in organisms with very high frequencies of exchange per bivalent, the simple calculation described above can seriously underestimate the actual map length. The simple crossover counting metric ignores the fact that bivalents with two crossover events can produce non-crossover, single crossover, and double crossover chromatids (see figure 7.9). Counting such single crossovers as resulting from only single exchange events will obviously underestimate the frequency of crossingover. For those interested

---

[4]   We urge readers to realize that any estimate of recombination frequency is just that, an estimate. The term *accuracy* has little or no value when applied to map lengths. The map length that you measure will be affected by the markers you use, by the genetic background, and perhaps even by environmental conditions. All of these factors will affect the *precision* of your estimate. If you are looking at some agent of genetic variation that might affect exchange frequency, it is wise to run a simultaneous control in an identical genetic background. If you are simply trying to map two or more variants, then the best you can probably get is a reasonable estimate.

in understanding or using a more complex mapping function that considers such issues, a more formal treatment is provided in the following paragraphs.

The mapping function

The difficulties in accounting for the effects of two or more exchanges within a given genetic interval were first considered by Haldane (1919). The key to that analysis is being able to use the crossover data; one has to predict the actual number of exchanges that occurred in a given interval. We can perform such an analysis under the following conditions:

1   The probability of exchange in a given interval is low.
2   The probability of two exchanges occurring in a given interval is simply the square of the probability of one exchange occurring in that interval. That is to say, there is no crossover interference.
3   There is no chromatid interference, that is if the choice of non-sister strands is random for each exchange. In this case, the ratio of double exchange types should be 1:2:1 for two-, three-, and four-strand doubles.

The first two assumptions allow us to estimate the number of single, double, triple, or greater exchanges that occurred in the interval of interest. The third assumption allows us to predict the types of crossover chromatids produced by those exchanges. For example, consider the case where two exchange events occur between a single pair of markers. From these various double exchange tetrads, we recover non-crossovers, single crossovers in each of the two regions of exchange, as well as the expected double crossovers:

2-strand doubles → 2 double crossovers; 2 non-crossovers
3-strand doubles → 1 non-crossover; 2 single crossovers; 1 double crossover
4-strand doubles → 4 single crossovers

The critical point to realize is that, without intervening markers, the so-called double crossover chromatids will not be recognizable as crossover events. That is they will not be recombinant for the two flanking markers. Thus, only the chromatids with a single crossover will be recognizable as "crossover chromatids." We could draw out the same classes of exchange bivalents for triple exchange tetrads, but we would obtain the same result. Half of the chromatids would be classifiable as crossover chromatids. (A mathematical justification for this empirical assertion can be found in the discussion of algebraic tetrad analysis below.) The bottom line is that whether a bivalent has one, two, three, or 20 crossover events occurring between two flanking markers, only half of the chromatids it produces will carry crossover chromatids.

Following Stahl (1969), we can now write an algebraic equation that relates the number of observable crossover events to the number of exchanges. This equation assumes that the number of exchanges per bivalent follows a Poisson distribution (i.e. there is no crossover interference). The average frequency of exchange is denoted by the symbol x. Thus, if the number of exchanges (or events) between any two markers is denoted by the symbol n, then the fraction of bivalents with no exchanges is:

$x^n e^{-x}/n!$

which for $n = 0$ reduces to e^{-x}. Thus the fraction of bivalents with one or more exchanges is defined as $1 - e^{-x}$. Because the frequency of observable crossovers will equal half the frequency of exchange bivalents, we can estimate the fraction of bivalents that will produce one or more observable crossover chromatids, denoted by the symbol P, by the equation:

$P = (1 - e^{-x})/2$

This equation has some lovely properties. One of them is that when x is small, P approaches $x/2$. This is really a restatement of the fact that the chiasma (or exchange) frequency equals half the map length.

We now wish to be able to describe the recombination fraction in terms of some real estimate of genetic distance, or map length, which takes into account the exchange events that do not produce crossover chromatids. This "real map length" is denoted by the value d. We will assume that d is proportional to P, and thus for closely linked intervals $d = x/2$ or $(x = 2d)$ and:

$P = (1 - e^{-2d})/2$

Again following Stahl (1969), we define a map distance as D, where $D = 100d$. (This is the equivalent of converting recombination frequencies to map units or centimorgans, multiplying by 100.) The equation then becomes:

$P = (1 - e^{-2D/100})/2$

This equation allows us to plot the observed fraction of recombinants (P) against the map length (D) in figure 7.10. Note that the observed recombination frequency increases linearly with the map distance over short distances. At longer distances the frequency of crossover chromatids will asymptotically approach 0.5. Thinking about this curve should allow us to see that map distances for adjacent intervals will be additive only for relatively small intervals. As the map lengths of the intervals increase, this additivity will become progressively worse.

As we noted above, this function presumes: (i) no sister chromatid exchange; (ii) no chromatid interference; and (iii) no crossover interference. The first two assumptions are fine, the third one is usually wrong. There is crossover interference in most organisms and thus double, triple, and multiple exchanges are less common than would be predicted by the Poisson distribution. This will have the effect of making the fraction of crossover chromatids (P) a more accurate approximation of genetic distance over larger genetic intervals.

How then does one determine the fraction of bivalents that underwent zero, one, two, or three crossovers, if the Poisson distribution isn't reliable? Indeed, it must be done empirically, using systems that allow you to recover all, or at least half, the chromatids produced by each meiosis and thus determine directly the number of exchange events per bivalent. This method is referred to as tetrad analysis, and is described below.

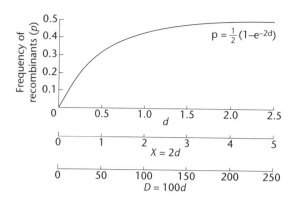

Figure 7.10 A mapping function for a hypothetical diploid organism that does not display crossover interference. The actual frequency with which recombinants are observed (p) increases linearly with map distance (D) over short distances. However, at longer distances the frequency with which recombinants are observed will asymptotically approach 0.5. See the text for a fuller explanation. (After Stahl 1969)

7.4.3 Determining the fraction of bivalents with zero, one, two, or more exchanges (tetrad analysis)

Why do we concern ourselves with converting crossover data into frequencies of tetrads with zero, one, two, etc. exchanges? What kind of information does the latter give us that is not apparent from the former? The answer lies in the fact that we are interested in the meiotic properties of chromosomes. The question of how chromosomes segregate is closely tied in with the nature of conjunction of bivalents, and we have seen that a chiasma can hold the elements of a bivalent in conjunction. When we consider questions of nondisjunction, or the properties of rearrangement heterozygotes, we have reason to consider the occurrence or non-occurrence of exchange. Thus we are often less interested in knowing the crossover frequency than in estimating the **exchange distribution**. The term exchange distribution refers to the actual distribution of chiasmata among the bivalents at meiosis I, i.e. the fraction of bivalents that underwent no exchange events (E_0), one exchange (E_1), two exchanges (E_2), and so on ($E_3 - E_n$).

Obviously, the best way of measuring the frequency of meiotic recombination (other than directly counting chiasmata) would be one that allowed us to recover all products of meiosis in a fashion that allows us to recover *all* four products of each meiosis, and thus to easily determine the exchange classification of the bivalent that produced those crossovers. This process is referred to as **tetrad analysis**. The advantage of certain classes of fungi (the ascomycetes, such as *Neurospora crassa*) for such tetrad analysis lies in the fact that all meiotic products are recovered in a linear ascus according to the sequence of segregations in the two meiotic divisions [for a review, see Perkins (1955, 1962)]. Alternatively, there is an algebraic method of estimating the exchange distribution that can be used when one can only recover single products of meiosis [Weinstein (1932); but see Zwick et al. (1999)]. We begin with a discussion of full tetrad analysis, then consider the analysis of half tetrads, and finally return to algebraic methods for single chromatid analysis.

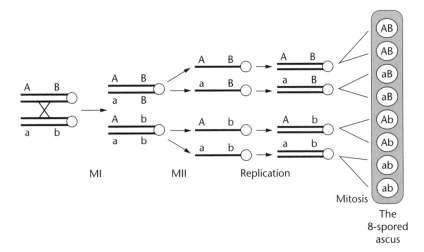

Figure 7.11   The Neurospora ascus: segregation of a centromeric marker (B) and a distal marker (A) with an exchange proximal to gene *A*

Tetrad analysis in linear asci in Neurospora

In Neurospora, the two meiotic divisions occur in such a way as to create a linear array of four meiotic products. The order of each doublet of cells in the ascus corresponds to the order of chromatids on the metaphase plate. The two cells above the dashed line arose from one product of MI and the two cells below the dashed line arose from the other (figure 7.11). Each of these four meiotic products then undergoes a complete mitosis to create two daughter cells. The linear order of spores in each ascus makes it possible to map the centromeres in relation to linked markers, according to the frequencies of first or second division segregation of markers.

If no exchange occurs between the **A** gene and its centromere, then the order of spores in the ascus will be **AAAAaaaa**. Now, suppose a single exchange occurred proximal to marker A. We might now expect the order of spores to be **AAaaAAaa**. (Other possible sequences are **aaAAaaAA**, **AAaaaaAA**, and **aaAAAAaa**, depending on the orientation of the two chromosomes on their MII spindles.) The critical issue here is that the **A** and **a** alleles did not separate into separate cells at MI. Rather, the occurrence of an exchange prevents the two different alleles from being separated into separate cells until meiosis II.

Using the terminology of fungal genetics, the absence of an exchange between a marker and its centromere is indicated by first division segregation and the presence of that exchange is heralded by second division segregation. (*Incidentally, the occurrence of second division segregation provides strong evidence that crossingover occurs between chromatids at the four-strand stage of meiosis I; crossingover between unreplicated chromosomes would not give rise to second division segregation.*) The frequency of asci showing second division segregation is a measure of the chiasma frequency; but only half the spores in each of those asci will carry crossover

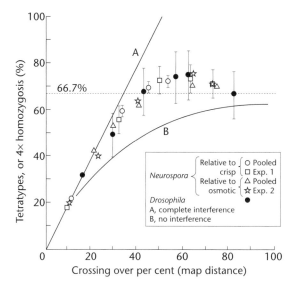

Figure 7.12 The Neurospora mapping function: interference as manifested in the frequency of tetratype segregations relative to the terminal markers *cr* or *os*. 95% confidence limits are indicated for critical points. (After Perkins 1962)

chromatids. Thus, we must divide the frequency of asci showing second division segregation by two to obtain the frequency of crossover chromatids.

Using several linked markers, we can confirm many of the rules of crossing-over listed previously and obtain more direct estimates of the exchange distribution. Most importantly, we can directly assess the frequency of the three classes of double crossover events. By confirming that the ratio of two-strand:three-strand:four-strand double exchanges was indeed 1:2:1, we can verify the assertion that there is no chromatid interference.

The existence of such double exchanges, and presumably higher order exchanges, creates an oddity in the expected relationship between map distance and the frequency of second division segregation. Naively one might expect that as one progresses away from the centromere the frequency of second division segregation would gradually increase to a maximum of 100%. Thus the maximum map length would be 50%, the equivalent of independent assortment. In reality, in Neurospora the frequency of second division segregation rises, but not linearly, to a maximum of just below 80%, and declines to approach a final value of 66.6% (figure 7.12) (Perkins 1962).

How then are we to understand the fact that the relationship of second division segregation to map length (as determined by summing small intervals) is not linear? And why does it eventually decline from values often as high as 80% (and for some organisms as high as 88%) to a final value of 66.6% for all genes studied? The answer to both questions lies in the fact that over long distances the double (and higher level) exchanges will effectively randomize the order of the four copies of each marker (gene) within a tetrad. In this case, the maximum frequency of second division segregation is 66.6%. If this is confusing just imagine that you had four spores, **AAaa**, in your hand like marbles and you are now going to drop them one at a time *at random* into a test tube-like ascus. If you drop the

A-bearing spore in first, then you have a two-thirds chance that the next spore will be an **a** spore (this will look like second division segregation, however the next two spores fall) and only a one-third chance that the second spore to fall will be an **A** spore (obviously, this will look like second division segregation, however the next two spores, both **a** spores, fall). For this reason the maximum possible frequency of second division segregation is going to be 66.6%, and thus the highest possible recombination frequency will be 33.3%.

However, we might also ask why the initial frequency of second division segregation can rise well above this eventual "maximum" of 66%. The answer lies in the fact that crossover interference is high in Neurospora (indeed, double crossing-over is virtually absent in intervals of less than 20 cm). Thus in the leftward part of the graph (centromere proximal) we are seeing only the effect of increasing the frequency of single crossover events. Based on the tenet that a single exchange may involve any two non-sister crossover chromatids, we expect that the maximum frequency of second division segregation will approach 100% (the case where all bivalents in the population have one exchange). (The alert reader may wonder why this "hump" in the mapping function was not observed in Stahl's idealized graph presented in figure 7.10. The answer lies in the fact that Stahl's idealized function was written for creatures lacking interference.)

The fact that we do see values in excess of 66% is critical, because it provides the strongest evidence that either sister chromatid exchanges do not occur, or if they do, they do not compete (or interfere) with inter-homolog exchanges. (If sister exchanges were as likely as non-sister exchanges, then the maximum possible frequency of second division segregation would be 66.6%, even in that mythical population of single exchange bivalents.) Perkins (1955, 1962) has tabulated multiple examples of frequencies of second division segregation that exceed 66.6% in a variety of fungi, including cases in Podospora that exceed 95%. These data, as well as the Drosophila half-tetrad data presented in box 7.4, clearly validate the assertion that either sister chromatid exchanges do not occur, or if they do, they do not interfere with inter-homolog exchanges.

An excellent review of the uses of ordered tetrad analysis may be found in Zhao and Speed (1998a) [see also Zhao et al. (1995)]. This article also contains a critical discussion of the statistical issues inherent in linear tetrad analysis. Although there a few other fungi that produce ordered asci, the most commonly used fungi (the yeasts) do not. Rather, these organisms produce asci with unordered tetrads. Although tetrad analysis in these organisms is a bit more difficult, it is none the less possible.

Unordered tetrad analysis in yeast

Two of the most commonly used organisms for genetic analysis, *S. cerevisiae* and *S. pombe*, also produce asci following meiosis. Each ascus contains four spores, with each spore consisting of a single meiotic product. Unfortunately, however, the spores are not ordered. They are actually arranged in a pyramid-like structure within the ascus. None the less, we can still do genetic analysis. We can no longer use the equator of a linear ascus as a centromere marker, but we can follow the segregation of any two other markers. Let us now extend our analysis to two genes (**A** and **B**), which may be located on the same or different chromosomes. Given that the initial mating consisted of haploid **AB** cells to haploid **ab** cells, the resulting diploid will have the genotype AB/ab. Following sporulation one can recover the following three classes of tetrads: parental ditypes (PDs), non-parental ditypes (NPDs), and tetratypes (TTs).

AB	**Ab**	**AB**
AB	**Ab**	**Ab**
ab	**aB**	**aB**
ab	**aB**	**ab**
Parental ditype	Non-parental ditype	Tetratype

Imagine that **A** and **B** are on different chromosomes, but so near their centromeres that recombination does not occur between either marker and its centromere. In that case we will observe only PDs and NPDs. (Convince yourself that you will see a tetratype *only* if recombination has occurred between at least one of the two markers and its centromere.) Because the two pairs of centromeres will orient randomly with respect to the other at meiosis I, PDs and NPDs will be equally frequent.

Now, suppose that the **A** and **B** genes were immediately adjacent to each other on the same chromosome. In the absence of exchange between them, all you would see is PDs. Rare exchanges between them could create tetratypes, but an NPD could only result from a presumably quite rare four-strand double crossover (figure 7.13). Thus in this case, the frequency of PDs will greatly exceed the frequency of NPDs. An excess of PD asci relative to NPDs is evidence for linkage. Actual estimates of map length can be determined by using the formula presented in box 7.5. The methods for mapping centromeres in unordered asci, and for mapping genes to either side of the centromere in these organisms, are presented in box 7.6.

Unordered tetrad analysis is also possible in a higher plant, *Arabidopsis thaliana*. The mutation *quartet1* (*qrt1*) causes the four products produced by each pollen meiosis to remain attached to each other, making tetrad analysis possible (Copenhaver et al. 1988, 2000). Studies using these mutants have produced an exceedingly thorough understanding of the control of exchange distribution in this organism. Similarly, tetrad analysis has been described and exploited in the green algae *C. reinhardtii* (Dutcher 1995).

Half tetrad analysis

Unfortunately, in most higher eukaryotes it is still not possible to perform full tetrad analysis. It is, however, possible in many organisms to recover two of the four products of meiosis. This was first accomplished by using compound (or attached) X chromosomes in Drosophila. This analysis was presented in box 7.3. Subsequent workers have developed tools for half tetrad analysis in a variety of

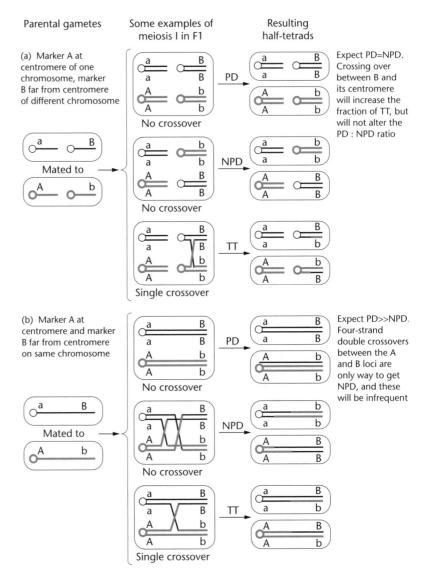

Figure 7.13 The effect of linkage on the recovery of PDs and NPDs. (Adapted from Johnson et al. 1995)

other organisms (including mice and humans) [for a review, see Johnson et al. (1995)].

Perhaps the most useful of these methods has been developed in a vertebrate system, namely zebrafish. Using a technique developed by Streisinger et al. (1986), meiotic half tetrads (derived from a single product of meiosis I) can be generated from females of virtually any genotype [cf. Johnson et al. (1995)]. This is accomplished by using high pressure to block the second meiotic division. Blocking meiosis II leads to diploid ova, which when fertilized by UV-inactivated

Box 7.5 Using tetrad analysis to determine linkage

We can use the three classes of asci to determine the map length between any pair of markers. In organisms, where interference is high, one could simply use the half frequency of TTs as a good estimate of map length. (Again the frequency of TTs is really a measure of the chiasma frequency, and is equal to twice the frequency of observed crossover chromatids.) But in most organisms this metric would underestimate the amount of recombination by failing to consider the double crossover events. Only a quarter of such doubles will generate NPD asci; three-strand double crossovers will generate TTs, while two-strand doubles are not observable. So we can estimate the frequency of doubles by 4(NPD). To account for the fact that each double exchange has *two* exchanges, we will count these twice, and thus the number of crossovers resulting from double exchanges is 8(NPD).

Unfortunately our TTs can also arise from three-strand doubles. We want our estimate of single exchanges to be free of these events. Fortunately, the frequency of three-strand doubles can be estimated as 2NPD. (Remember the ratio of two-strand:three-strand:four-strand doubles is 1:2:1 and the frequency of NPDs is the frequency of

four-strand doubles.) Thus the true frequency of single exchange is equal to TT − 2NPD.

We can now sum the frequency of exchanges involved in both single and double crossover events by adding these two quantities: (TT − 2NPD) + 8NPD = TT + 6NPD. But since each exchange produces two crossover and two non-crossover chromatids, we need to divide this quantity by two to get the frequency with which we recover crossover chromatids (the map length). Thus:

$$\text{Recombination frequency} = \tfrac{1}{2}(TT + 6NPD)/(PD + TT + NPD)$$

or

$$\text{Recombination frequency} = (TT + 3NPD)/\text{Total tetrads examined}$$

Map length is expressed as 100 times the recombination frequency, and thus:

$$\text{Map length} = [(TT + 3NPD)/\text{Total tetrads examined}] \times 100$$

This is the so-called Perkins equation and can be used in organisms with either ordered or unordered tetrads.

Box 7.6 Mapping centromeres in fungi with unordered tetrads

Mapping centromeres

The methods described in the text allow any two genes defined by mutant alleles to be mapped relative to each other by unordered tetrad analysis. But how can one map the centromeres by this method? The answer is, in pairs. Two markers that both define their centromeres will show an equal frequency of PDs and NPDs without producing tetratypes. Thus once one has identified one centromere marker (and this was initially done by Don Hawthorne using a variant of *S. cerevisiae* that produced linear, but oddly ordered, asci) one can easily find others.

Mapping two genes that are very close to the same centromere

How might you map genes to opposite sides of the centromere in an organism like yeast that yields unordered tetrads? Consider the cases **ABD**/**abd** where **A** and **B** are tightly linked and where **D** is a very tightly linked centromere marker on another chromosome. Consider the pair [**AD** or **BD**] that gives the lowest frequency of tetratypes. Say that pair is **AD**. Among those AD tetratypes, **B** will segregate with **A** if they lie on the same side of the centromere (i.e. all of these tetratypes will be PD with respect to **A** and **B**). But if the **A** and **B** are on opposite sides of the centromere, then these same **AD** tetrads will be TT with respect to **A** and **B**.

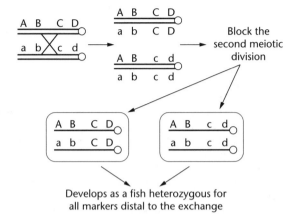

Figure 7.14 Zebrafish half tetrads

Develops as a fish heterozygous for all markers distal to the exchange

sperm develop as gynogenetic, or matroclinous, females. Such females will always be homozygous (or reduced) for centromeric markers. Indeed, in the absence of any exchange they would be homozygous for all markers. But the occurrence of a single exchange in the preceding meiotic prophase will result in heterozygosity for all markers distal to the exchange (figure 7.14).

The distance between any gene and its centromere can then be calculated as half the frequency with which heterozygous gynogenetic offspring are recovered from a cross involving females heterozygous for a recessive allele at that locus.[5] Unfortunately, heterozygotes are often hard to distinguish from wildtype homozygotes. But we can estimate the frequency of wildtype homozygotes by using the easily assayed fraction of mutant homozygotes. Thus one can do the calculation by subtracting twice the frequency of mutant homozygotes from one and dividing that number by two. This calculation, which presumes absolute interference, has been shown to work well for short gene to centromere distances. Methods for the statistical analysis of half-tetrad data have been presented by Zhao and Speed (1998b) [see also Zhao et al. (1995)].

Johnson et al. (1995) have presented a more general application of this method for linkage analysis that allows the investigator to determine gene–centromere distances even over distances large enough to have allowed double crossover events. The method is based on following the segregation of a given marker with respect to the segregation of a PCR-based centromere marker. As shown in figure 7.13, if the two markers lie on the same chromosome then PD>>NPD. If they are unlinked, then PD = NPD. Once markers have been positioned on a given linkage group, the distance between them can be assessed by determining the fraction of time the two markers differ in terms of first versus second division segregation. (That is, how often offspring are produced that remain homozygous for the proximal marker a, but are now heterozygous for the more distal marker b. Such progeny must result from exchange between the

---

[5]   Please tell us that you understand why we multiplied by 0.5 here. Remember that each animal carries two chromatids, only one of which was crossover-bearing.

two markers.) This method has allowed the construction of an extremely detailed map in zebrafish.

Weinstein's algebraic tetrad analysis

Unfortunately in most organisms it is still not possible to use either full or half tetrad analysis to determine the exchange distribution. However, even in cases where we can only recover one product of meiosis, there is an algebraic method to determine the tetrad distribution that produced a given set of meiotic products. That method was developed and applied to Drosophila crossover data by Alexander Weinstein (1932). Our own explanation of the method is provided below.

In the trivial case, where there is only one homologous crossover in a tetrad, it should be clear that two (and only two) of four chromatids are involved. This situation produces two chromatids that have a crossover, and two that do not. The situation gets somewhat more complicated when we begin to analyze tetrads with more than two chromatids involved in crossingover. As we have noted previously, there are three types of double exchanges and the ratio of two-strand: three-strand:four-strand doubles is 1:2:1. Similarly, the classes of crossover events recovered from these double exchanges, non-crossovers, single crossovers, and double crossovers, are also produced with a ratio of 1:2:1. This distribution, 1:2:1, is a very simple form of a binomial distribution. For tetrads with increasing numbers of exchanges, the frequencies of chromatids recovered in each class continue to follow a binomial distribution. Thus by the binomial expansion theorem, in a given tetrad with n exchanges, the frequency of chromatids with k crossovers will be:

$$K = [n(n-1)(n-2) \ldots (n-k+1)/k(k-1)(k-2) \ldots 1](0.5^n)$$

(For an in-depth treatment of the binomial coefficient and its derivation, refer to any sufficiently advanced text on mathematical analysis.)

We can denote the state of each bivalent as an **exchange rank**. A bivalent with exchange rank equal to zero is simply a bivalent with zero crossovers. A bivalent with exchange rank equal to one has one crossover, etc. We can also denote the number of crossovers on chromatids as follows: NCO, non-crossover; SCO, single crossover; DCO, double crossover; TCO, triple crossover; and so on. Using this notation, and considering all possible tetrads, we could calculate as follows:

Exchange rank	Fraction of strands recovered as:				
	NCO	SCO	DCO	TCO	QCO
0	1	–	–	–	–
1	1/2	1/2	–	–	–
2	1/4	1/2	1/4	–	–
3	1/8	3/8	3/8	1/8	–
4	1/16	4/16	6/16	4/16	1/16

(Those with a mathematical background may recognize this as a simple modification of Pascal's triangle. This is no accident, since both are derived from the binomial coefficients.)

Keep in mind that when we count crossovers, we have no definitive way of establishing that a given non-crossover was derived from a no-exchange tetrad, or that a chromatid with a single crossover came from a tetrad with one exchange, etc. Unfortunately, we cannot simply examine a living fly and determine the classes of tetrads that produced its chromosomes. In order to determine the number of exchanges that occurred in a tetrad, we must approximate it from recombination data, which we can observe by marker exchange. In essence, we must work backward, deriving the number of tetrads falling into each class by observing the relative frequencies of each class of recombinant. First we will examine a simple case, and then attempt to provide insight into how this would be mathematically generalized. In order to transform recombination data into tetrad frequencies, we must introduce some new variables:

a_n = the frequency of chromatids with n crossovers (e.g. a_2 is the frequency of double crossovers).

E_n = the frequency of tetrads with n exchanges. [Note that E is a frequency, which may be confusing at times due to the fact that people (ourselves included) will simply refer to a tetrad with n exchanges as "an E_n."]

For the sake of simplicity in the following example, we will make the assumption that only tetrads with zero, one, or two exchanges will occur. In the context of this admittedly contrived example, all of the observed double crossover chromatids observed must come from tetrads in which two exchanges have occurred. Furthermore, the frequency of tetrads having two exchanges will directly affect the number of double crossover chromatids that we observe. (If E_2 is very low, there is little opportunity for a double crossover to form, however if it is high, double crossovers are far more likely.) Even though we don't know the value of E_2, we are able to predict which fraction of these double exchange tetrads will give rise to double crossover chromatids. Remember that in a given tetrad with two exchanges, when sister chromatid crossing over does not occur, the frequency of double crossover chromatids produced is 0.25. Thus in our example, the total frequency of double crossovers should be 0.25 multiplied by the frequency of tetrads with two crossovers. Mathematically this is written:

$$a_2 = (0.5)^2 E_2 = 0.25(E_2)$$

The case of single crossover chromatids is a little bit more complicated. Single crossover chromatids can be produced by tetrads with either one or two exchanges. Again, half the chromatids coming from tetrads with two exchanges should be single crossover chromatids. Similarly, half the chromatids from tetrads with one exchange should have one crossover. Since these two types of tetrads are the only source of single recombinants in our example, we can add these two values together to give the total value for chromatids with one crossover:

$$a_1 = 0.5(E_1) + 0.5(E_2)$$

Finally, in the case for chromatids with no crossovers, we must consider that all three classes of tetrad are producing them:

$a_0 = E_0 + (0.5)E_1 + (0.5)^2 E_2 = E_0 + 0.5(E_1) + 0.25(E_2)$

Now here are our three equations:

$a_0 = E_0 + 0.5(E_1) + 0.25(E_2)$
$a_1 = 0.5(E_1) + 0.5(E_2)$
$a_2 = 0.25(E_2)$

We can then use simple algebra to solve this system for the values of E:

$E_2 = 4a_2$
$E_1 = 2a_1 - 4a_2$
$E_0 = a_0 - a_1 + a_2$

We now have three equations, describing the relative frequency of each class of tetrad, in terms of values of a_n, which we simply determine experimentally. Of course, as we stated previously, this example is far from realistic. In the real world, we cannot make assumptions such as "no tetrads with more than two crossovers." The previous example does however provide the idea of how actual tetrad analysis is done. To generalize the previous equations, we write out the initial equations for a_n in terms of E_n, with our terms still being derived from the binomial coefficients. It looks like this:

$a_0 = E_0 + 0.5(E_1) + 0.25(E_2) + 0.125(E_3) + 0.0625(E_4) + \ldots + P_{n,0}(0.5^n)E_n$
$a_1 = 0.5(E_1) + 0.5(E_2) + 0.375(E_3) + 0.25(E_4) + \ldots + P_{n,1}(0.5^n)E_n$
$a_2 = 0.25(E_2) + 0.375(E_3) + 0.375(E_4) + \ldots + P_{n,2}(0.5^n)E_n$
$a_3 = 0.125(E_3) + 0.25(E_4) + \ldots + P_{n,3}(0.5^n)E_n$
$a_4 = 0.0625(E_4) + \ldots + P_{n,4}(0.5^n)E_n$

In general:

$a_m = P_{m,m}(0.5^m)E_m + P_{m+1,m}(0.5^{m+1})E_{m+1} + P_{m+2,m}(0.5^{m+2})E_{m+2} + \ldots + P_{n,m}(0.5^n)E_n$

Here, up to n crossovers are observed. The value $P_{n,k}$ is given by the binomial coefficient formula (Pascal's formula):

$[n(n-1)(n-2) \ldots (n-k+1)]/[k(k-1)(k-2) \ldots 1]$

Although these equations are clearly far more complicated than the ones in the previous example, deriving them requires the same application of binomial coefficients, and solving them relies only on the same basic algebraic principles that we have already employed. For an in-depth treatment of the solution of these equations, see Weinstein (1932). When these equations are solved, our final equation for estimating the frequency of a tetrad with r crossovers and no sister chromatid exchange becomes:

$E_r = 2^r[a_r - a_{(r+1)} + a_{(r+2)} - a_{(r+3)} + \ldots + (-1)^n a_n]$

Although Weinstein's method has been successfully applied in several organisms, it does suffer from some exceedingly irrational limitations. Most notably, it often returns negative tetrad frequency estimates that are biologically meaningless, especially for the frequency of non-exchange tetrads. Zwick and his collaborators have recently published a modification of the Weinstein approach, which bypasses this difficulty (Zwick et al.1999). It is also possible to interpret Weinstein's tetrad analysis data in terms of the position of exchanges along the arms of meiotic chromosomes. Szauter (1984) and Carpenter (1988) have considered this technique, referred to as regional tetrad analysis, in detail.[6]

7.4.4   *Statistical estimation of recombination frequencies (LOD scores)*

The human geneticist mapping human genetic loci (locations of genes, markers, phenotypes, etc.) is often faced with using statistical tools to estimate the map distance between a given set of markers or between a marker and a locus. In such cases the available data may sometimes come from large families showing simple Mendelian inheritance, but frequently such studies pool information from multiple small families, or even groups of affected relative pairs. In traditional forms of linkage analysis, known as parametric forms of analysis, assumptions about the mode of inheritance are included in the evaluation of the results. In other non-parametric forms of analysis (for example analytical approaches used in studying pooled data on sibling pairs) the calculations are carried out in a way that allows for evaluation of the data even if your assumptions about mode of inheritance are wrong or non-existent.

Two-point linkage analysis

The traditional form of linkage analysis, called a two-point analysis, evaluates apparent linkage between two points in the genome. Frequently, when attempting to map a phenotype being transmitted in a family, what is being evaluated is linkage of a single marker (which can be a DNA marker, a blood type, or other marker) and a locus (a position in the genome that is responsible for the phenotype that is being mapped). Conceptually, what takes place is a test of the hypothesis that two items are linked and will be transmitted together between generations more frequently than would happen if they were unlinked. We can consider such a test of linkage for two different traits (blond hair and blue eyes), for two different DNA-based markers (D1S210 and D1S452), for two non-DNA markers (a blood group and a tissue transplantation antigen), for two different disease phenotypes (e.g. aniridia and Wilms tumor), or for any combination of two items from this list. However, if the geneticist is after the location of a disease

---

[6]   As previously noted, all evidence points toward a negligible level of sister chromatid exchange *in vivo*, in most organisms. If sister chromatid exchange were a factor, however, it would change our equations for doing tetrad analysis. For example, in the presence of sister chromatid exchange, a tetrad with one crossover no longer definitively yields two recombinant chromatids and two non-recombinants, but could instead give rise to four chromatids, which appear to be non-recombinant. Because sister chromatids are copies of each other, there is no way to observe sister chromatid exchange using conventional marker recombination. While all of the previous equations would need to be modified to account for sister chromatid exchange, the basic principles would be the same.

gene then what will usually be tested is the hypothesis of linkage between a particular allele of a genetic marker (these days that would be a DNA-based marker such as a microsatellite repeat marker) and the presence of the disease.

One of the key pieces of information that tells us whether the two items are linked is the apparent distance between the two items. An allele of a marker that is located dead on top of the locus being mapped will always co-segregate with the locus, giving an apparent marker–locus distance of zero. A marker and a locus that are located on different chromosomes (or far apart on the same chromosome) will segregate independently, as shown by approximately 50% recombination between the marker and the locus. The distance between the marker allele and the locus is called the recombination fraction or theta (θ). Most of the time we don't know what the real distance is, but can only estimate what the distance appears to be based on the small slice of information we can see in this particular family or data set.

We need to clearly state here that the value of θ we arrive at by the time we are done is a theoretical value rather than a real value, so we put the symbol $\wedge$ above the θ to indicate our best estimate of the real value. In carrying out linkage analysis, we will test many possible values of θ in an effort to determine which value of $\hat{\theta}$ is closest to the real (but unknown) value of θ.

To obtain the best estimate of which value of θ is "right," we need to know which values of θ are most consistent with the data that we have in hand. This will require a set of statistical tests. These tests are designed to "model" the inheritance of two markers for any given value of theta. Using that model we can ask how likely it would be to obtain a given data set for that value of theta. This model can then be compared to other possible models generated using different values of theta. Such a comparison allows us to determine the best model (the one that gives the best fit to the data) and determine a statistical likelihood that any given model is correct.

The key piece of information we need if we are going to assess the significance of any particular value of θ, or the likelihood that it is the best fit to the observed data, is a measure called the LOD score (indicated by the symbol Z). As we test each value of θ across the range from 0.00 (complete linkage) to 0.50 (unlinked), we obtain a LOD score for each value of θ. For any particular marker we end up considering the value of θ that gives us the highest LOD score (also sometimes called maxLOD) to be the most statistically significant indicator of the real value of θ. Thus, what we actually do is test a range of values across the entire interval of possible values (from completely linked to unlinked) and select the value of θ that is associated with the highest LOD score $\hat{Z}$ for that particular marker.

The term LOD can be thought of as coming from the phrase "log odds" or "log of the odds ratio." The odds ratio effectively compares the probability that the two items are linked to the probability that they are unlinked. Taking the $\log_{10}$ of this ratio then provides values that are easier to talk about and compare than the values resulting from the actual odds ratio calculation, which can sometimes get to be very large and cumbersome to deal with. One of the big advantages of dealing with logarithmic values instead of the odds ratio itself is that when data on several different families are obtained, the LOD scores can simply be added together to give a combined LOD score.

$$LOD = \log_{10} \left(\frac{\text{probability of obtaining the observed data if the two genes are linked with a recombination frequency } \theta}{\text{probability of obtaining the observed data if the two genes are unliked } (\theta = 0.5)} \right) \qquad (7.1)$$

The odds ratio in this case compares the odds (or probability) that the marker and locus are linked (close together on the same chromosome) to the odds that the marker and locus are unlinked (on separate chromosomes or far apart on the same chromosome). In doing this comparison, we have to test specific distances between the items, such as testing for whether they are so close together that 1% of the meiotic events show recombination ($\theta = 0.01$) or linked a bit less closely so that 20% of the meiotic events show recombination ($\theta = 0.20$).

If there are no recombination events observed, then the observed value for θ is taken to be zero. If $\theta = 0.00$ then the LOD score can be calculated from the simple formula:

$$Z = n \log (2) \qquad (7.2)$$

where n is the number of informative meioses in that family. Interestingly, one of the things we can see from this equation is that in the case where there are no recombination events, each individual contributes about 0.301 (that is to say, $\log_{10} 2$) to the LOD score.

If the recombination fraction is a value other than zero, then the LOD score is calculated using the formula:

$$Z = n \log (2) + k \log (\theta) + (n - k) \log (1 - \theta) \qquad (7.3)$$

where n is the number of informative meioses and k is the number of recombinant individuals. When we look at equations (7.2) and (7.3), we see that the LOD scores will potentially be quite different if we test two individuals rather than 20 or 200. The LOD scores will be potentially higher when more of the tested meioses are informative. The LOD scores will be lower if there are more recombinants. In fact, when two families of the same size are compared, the one with more recombinants will have the lower LOD score.

These equations really apply only to the simplest situation of autosomal dominant inheritance in a two-generation family (parents plus children), in which we have complete information available for both markers and phenotypes for every single person screened. As soon as any of those conditions change, including a more complex family structure, missing information, or a different mode of inheritance, the computations become much more complex. Most linkage calculations these days are carried out using computer programs. Those who are interested in learning more about LOD score calculations in more complex circumstances could consult Ott (2001).

But what about this phrase "informative meioses?" The number of people we test is not actually the critical determinant of the value of n. What actually

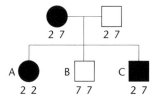

Figure 7.15   A partially uninformative family. Filled symbols are affected individuals and open symbols are unaffected. Individuals A and B are both informative because we can tell which allele they received from each parent and specifically we can tell which allele they received from the affected mother. Individual C is uninformative because we cannot tell whether his mother passed him allele 2 or allele 7

determines *n* is how many of the people tested gave us information that told us whether they were recombinant or non-recombinant, which requires that we be able to tell which of two copies of the marker got passed along from the affected parent. Two neighboring markers can give very different LOD scores for the same recombination fraction if one of the markers is less informative.

An example of an uninformative situation would be a family in which both parents and all children were homozygous for the same allele of a marker. (Notice that most of the markers that actually get used have been selected to have properties that should avoid this eventuality, but it still happens from time to time.) One of the other situations in which reduced information is obtained happens when both parents are heterozygous for the same alleles, in which case they can produce heterozygous (uninformative) children in addition to homozygous (informative) children. To see how this works look at figure 7.15 to see that child A received allele 2 from her father and allele 2 from her affected mother, which makes her informative since we can tell which allele the affected parent gave her. Child B received allele 7 from his affected mother and allele 7 from his father, once again informative. Child C received one copy of allele 2 and one copy of allele 7, but we cannot tell whether allele 2 came from the affected mother or from the father. The same situation holds for allele 7. The result is that we cannot tell which allele the affected mother passed to child C. So, even though we tested three children in this family, only two of them are informative. Thus this marker will produce a lower LOD score than a different marker that is fully informative in all three children.

Unfortunately, unlike those of us lucky enough to be working with flies or yeast, the human geneticist cannot always simply decide to generate larger numbers. Geneticists studying human beings also face other limitations to their experimental design that are not present in most of the model systems being used in genetics. One of the most common problems in human pedigrees is missing information. In some cases this is because members of the pedigree are deceased, including in some cases critical relatives who connect some members of the pedigree to the rest of the pedigree. In other cases, a family member may decide that they do not want to participate in the research project. In these cases, the computations become more complex. Among other things, in cases where someone is not available for screening with genetic markers but clearly is still a part of the pedigree structure, part of the computational process attempts to recreate what the genotype of that person might have been. In some cases, more than one alternative genotype is possible, and the computation must be retried with each of the possible genotypes considered. In such cases, one of the key pieces of information that is often needed will be the frequency of the different alleles in the population. This matters since an allele that is very common in the population will be much more likely to be part of the missing person's genotype than an allele that is rare. One of the other key issues here is that there are many genetic markers for which the allele frequencies are different for different racial and ethnic groups, so it is important to know whether the allele frequencies being considered have been derived from a relevant control population.

In other cases, the missing piece of information is a piece of medical information needed to confirm the disease phenotype, such as in cases where a relative has not had some specialty test done that would provide the conclusive diagnosis, or when someone died before they were old enough to manifest the trait. In many cases, pieces of medical information end up being missing for spouses who marry into a family but are not in the line of descent. Can we simply consider them to be unaffected and not at risk? No, they actually also have a level of risk that can be evaluated from our knowledge about the frequency of the disorder in the population. In these cases, one of the additional complications to the LOD score computation is the need to know something about how common the disease is in the population, and once again it is important that the disease frequencies being used come from a population of similar origins to the person being "reconstructed."

What constitutes statistically significant evidence for linkage?

The standard in the field is that a LOD score of 3.0 or greater is taken as providing highly significant evidence in favor of linkage of two items at a distance indicated by the value of θ associated with that LOD score. However, when a very large number of tests for linkage are carried out, such as when a very closely spaced genome scan is done, issues of multi-testing arise (errors of multi-testing refers to the probability of obtaining a spurious result just because one is testing so many markers.). In such cases a LOD score higher than 3.0 might be required. On the other hand, in other cases a lower LOD score may be allowed as indicating significant evidence of linkage. For instance, for a disease that is already known to be X-linked prior to carrying out any marker testing, a LOD score lower than 3.0 may sometimes be taken as an indication of significant evidence of linkage to a marker on the X chromosome.

Because the LOD score is $\log_{10}$ of the odds ratio, a LOD score of 3.0 indicates 1,000 to 1 odds in favor of linkage. The standard in the field for determining that two things are not linked is that a LOD score of -2.0 or lower indicates statistically significant evidence against linkage. A LOD score of -2.0 translates to odds of 100 to 1 against linkage.

Table 7.1   Two-point analysis of linkage between chromosome 1 markers and an eye disease gene in a single large kindred

Locus	Recombination fraction						$\hat{\theta}$	$\hat{Z}$
	0.01	0.05	0.10	0.20	0.30	0.40		
D1S194	−0.49	0.09	0.24	0.25	0.14	0.01	0.15	0.27
D1S196	1.90	2.34	2.31	1.90	1.29	0.56	0.07	2.36
D1S433	2.62	2.41	2.14	1.58	1.01	0.45	0.00	2.68
D1S452	6.32	5.88	5.30	4.07	2.70	1.22	0.00	6.42
D1S210	3.12	2.88	2.56	1.89	1.20	0.50	**0.00**	**3.18**
D1S1619	2.67	2.47	2.22	1.69	1.10	0.48	0.00	2.71
D1S2634	0.31	0.29	0.27	0.21	0.15	0.08	0.00	0.32
D1S242	3.73	4.04	3.83	3.04	2.04	0.91	0.04	4.04
D1S215	−0.01	1.69	2.13	2.06	1.51	0.73	0.14	2.19

As an example of LOD score analysis, table 7.1 (kindly provided by Dr. Julia Richards at the University of Michigan) presents the results of a LOD analysis for linkage between various markers on chromosome 1 and an eye disease. In such an analysis the test for linkage evaluates one marker at a time without taking into account any information from the surrounding markers. Although the table shows values of θ across the top at 0.01, 0.10, 0.20, 0.30, and 0.40, the actual calculations tested a full range of values from zero to 0.5, and not just the sample values shown in the table.

When we look at table 7.1, we see a large amount of information and are faced with the problem of identifying where the important pieces of information are located. One of the important points to realize about LOD score tables is that for most purposes the important information is usually in the columns on the right-hand side of the table (here highlighted in gray). The recombination fraction and the LOD score values shown in most of the columns are just examples, but the values shown in the right-hand pair of columns represent the maximum LOD score obtained for that marker, $\hat{Z}$, and the recombination fraction, $\hat{\theta}$, that corresponds to that maximized LOD score and represents our best estimate of the real value of θ for that marker. In this case, if we want to know which marker provided the best evidence for linkage, we would conclude that it is D1S452, for which the critical pieces of information are underlined. Why do we pick D1S452? If we look in those two right-most columns we find that D1S452 gave $\hat{\theta}$ of 0.00 (no recombination) corresponding to $\hat{Z}$ of 6.42 (more than a million to one odds in favor of linkage), which is the highest LOD score of any of the markers listed.

Now it is tempting to say that this means D1S452 is closest to the locus out of all the markers in the table. That is not actually valid. D1S452 provides the strongest evidence in favor of linkage, but technically any of the genes in that same interval that show $\hat{\theta}$ of 0.00 with a high LOD score could potentially be the closest to the gene. In fact, several markers below D1S452 in table 7.1 turned out to be closer once the actual gene was cloned. How could D1S1619 ($\hat{Z}$ of 2.71) or even more amazingly D1S2634 ($\hat{Z}$ of 0.32) be closer to the gene we are looking for than D1S452 if they end up with such low LOD scores? The answer is that the LOD score is affected by several things, including not only the recombination fraction but also the number of informative meioses. Consider this. What if all the progeny in the family were informative for marker D1S452, some of the progeny turned out to provide uninformative information for D1S1619, and almost all of them were uninformative for D1S2634? Then we would see just the kind of thing we see in table 7.1, a highest LOD score for the most informative marker when the other factor, distance, is about equal at the level of resolution of this analysis. If you go back and review equation (7.2), you will find that n represents not how many people are in the family but rather how many people give you information that lets you tell whether or not there was a recombination event.

Multipoint LOD analysis

A multipoint LOD score is calculated in a way that takes information from several adjoining markers into account when evaluating the theoretical separation between two items being mapped. This analysis will also produce a LOD score, but the analysis may come out with its maximized LOD score at a recombination fraction near to but not exactly the value of theta for which the two-point LOD score maximized. Results of such a multipoint analysis may sometimes be displayed graphically, and make the probable location of the locus very easy to visualize. As shown in figure 7.16, the bottom of the graph marks the positions of the various markers along the chromosome relative to the estimated most likely position of the locus, which is placed in the center at the position marked with zero. As you can see, the highest LOD scores on the graph occur near the center, where the highest point on the graph indicates the highest LOD score (the scale on the left side of the graph). Marker 3 provides the highest multipoint LOD score, but only slightly higher than that found for several markers on either side. This kind of multipoint graph is usually easy to interpret: look for the mountain peak and you have the location of the marker that is providing the strongest evidence for linkage.

Local mapping via haplotype analysis

Once a gene has been mapped to a particular region of a chromosome, studies of additional families and more markers in the immediate vicinity of the locus can assist with narrowing the region thought to contain the gene that is being sought. One of the other multipoint approaches to evaluating co-transmission of human marker information for markers located close together on the same chromosome is through haplotype analysis. By displaying markers geographically arrayed as they are located on the chromosomes, we can evaluate in more detail which alleles are located on the same piece of DNA together, and likely to be transmitted together to the next generation.

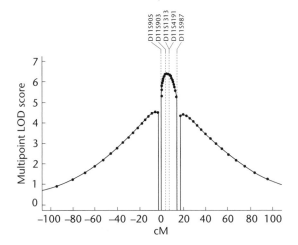

Figure 7.16   Multipoint linkage analysis of markers immediately around the NNO1 locus. Dots represent locations for which multipoint LOD scores were calculated to determine linkage between nanophthalmos and the marker map. D11S905-3.8 cM-D11S903-3.8cM-S11S1313-0.1 cM-D11S4191-7.0cM-D11S987. Dotted lines indicate the positions of these markers. Only one dotted line is shown for D11S1313 and D11S4191, because they are so close together. LOD scores of −∞ at D11S905 and D11S987 are not shown. (After Othman et al. 1998)

We are conveniently able to represent the alleles at multiple markers along a chromosome by placing the alleles in a column in the order in which the markers occur along the chromosome (figure 7.17). In this family, we want to be able to identify which alleles are being transmitted along with the disease, and determine which alleles are present together and transmitted together on a piece of DNA that contains the disease gene. As we can see in figure 7.17, affected individuals always have allele 9 at marker 3, allele 6 at marker 4, and allele 2 at marker 5, but affected individuals never have allele 4 at marker 3, allele 1 at marker 4, or allele 1 at marker 5. This suggests to us that there is a region of contiguous DNA containing alleles 9, 6, and 2 on one chromosome and alleles 4, 1, and 1 on the other. The affected father has passed along an "affected haplotype" (a linked set of alleles characteristic of chromosomes passed to affected individuals) consisting of alleles 9-6-2 at markers 4, 5, and 6 to his affected children and an "unaffected haplotype" consisting of alleles 4-1-1 at those same markers to his unaffected children.

In addition to letting us identify an affected haplotype, this kind of analysis allows us to see where recombination events have taken place. In figure 7.17, the right-hand chromosome in child A was produced by a recombination event during paternal meiosis between marker 5 and marker 6, which leaves child A carrying "unaffected" alleles at markers 5 and 6. The right-hand chromosome in child H was produced by recombination between marker 2 and marker 3 during paternal meiosis, which leaves child H carrying "unaffected" alleles at markers 1 and 2. This suggests to us that the disease gene should be located between marker 2 and marker 5, in the region that contains the "affected" haplotype.

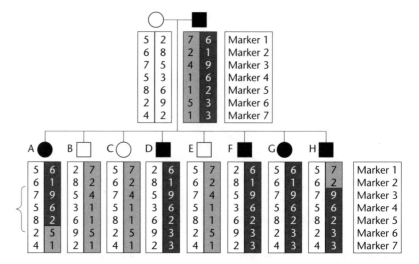

Figure 7.17 Haplotype analysis of child A and other children. Circles and squares denote the people in the family. Beneath each circle or square are two rectangles that diagram the two copies of the chromosome present in that individual with the markers arranged along the haplotype rectangle in the same order in which they sit on the chromosome. Filled symbols indicate individuals affected with the disease that has passed from one generation to the next in this family, in which the family history indicates autosomal dominant inheritance. Black rectangles with white letters show the alleles that co-segregate completely with the disease phenotype. Gray rectangles with black letters mark the other chromosome present in the affected parent. White rectangles with black letters mark regions of chromosome inherited from the unaffected parent. The bracket frames the region most likely to contain the disease gene. The affected haplotype is defined as a region of adjoining markers that are present on the same piece of DNA and transmitted together to all of the affected members of the next generation. In this case alleles 9-6-2 at markers 3, 4, and 5

When comparing information on multiple different families with the same disease, it is possible to compare the allele sizes found in the different families. In some cases it is possible to identify an affected haplotype from one family that turns up in another family with the same disease. This constitutes evidence that the families might share a common ancestor, although it is not proof. It is important to be careful when deciding that a shared haplotype can be interpreted as meaning shared ancestry. Consider that if a very small number of markers define the affected haplotype and the alleles in that haplotype are very common alleles, there should be concern that perhaps this is simply a fairly common haplotype present in many individuals in the population. This can be resolved by doing two things: looking at haplotypes in the general population to see whether the haplotype is present in individuals and families that lack the disease phenotype, and looking at additional markers in the same region in the two families being studied to see whether additional markers spaced in between the original markers continue to show allele-sharing between the two families.

The end game

Once the chromosomal region containing the gene has been identified (sometimes called the genetic inclusion interval or the critical region), the identification of the specific gene makes a great deal of use of informatics these days. As the draft version of the human genome sequence has gradually assembled itself toward a relatively complete sequence, a great deal of information on locations of human genes (as distinguished from human loci) has become available. Because the genetic markers are pieces of sequence whose positions are known within the human genome sequence, it is possible to first map a gene, identifying the flanking markers whose recombination events define the boundaries of the genetic inclusion interval, and then look that interval up in the database containing the human genome sequence. It is then possible to evaluate all of the genes listed within that interval as possible candidate disease genes. Types of information used to prioritize candidate genes to decide which ones to screen first include information on which tissues or cell types express the gene, information on what stages in development express the gene, and information on gene product function, biochemical pathway, or gene family. This process of prioritizing the candidate genes is important, since there can often be hundreds or even thousands of genes located within a genetic inclusion interval under consideration.

Once the candidate genes have been prioritized, then it is a simple matter of carrying out mutation screening to determine which genes in the interval contain sequence variants. You might think at that point it would be easy, but the expectation is that more than one gene in the interval will have sequence variants present, and only one of them will be the culprit causing the disease. How can we tell when we find the real disease gene? One of the things we can do is test to see whether the mutation is present in a population of people who do not have the disease. Another thing we can do is look for mutations in that particular gene in other families or individuals with the same disease. The formal proof of principle normally calls for doing something like showing a change in a measurable biochemical function associated with that mutation and gene, or constructing a transgenic animal that has the mutation and showing that the animal then develops the disease characteristics. This final step can be complicated on many levels, only one of which centers around the terribly obvious observation that mice are not humans and thus will not always develop the same phenotype in response to mutations in a particular gene. However, as a general rule so far, this approach has worked well. Map the gene, define boundaries to the genetic inclusion interval, reduce the size of the interval, identify the genes within the interval, prioritize them based on expression, function, and sequence homology information, test for mutations, evaluate other populations for the presence of those mutations, look for co-segregation of mutation and disease phenotype in other families, and get confirmation in an animal model or through biochemical testing.

Resources that offer additional information on linkage analysis are available and may be of use to the interested reader (Lander & Botstein 1989; Taylor et al. 1997; Ott 2001; Rice et al. 2001). Additional information on other methods of mutant mapping in human genetics can be found by looking at works on linkage disequilibrium [cf. Reich et al. (2001)], sib-pair analysis [cf. Farral (1997)], and population genetics [cf. Weiss and Clark (2002)].

7.4.5   *The actual distribution of exchange events*

The actual distribution of exchanges is not random among the bivalents that comprise a karyotype; the distribution of the number of chiasma per bivalent is much narrower than that predicted by a Poisson distribution [for reviews, see Jones (1984)], so there are almost no achiasmate bivalents and few with high numbers of chiasmata. For example in rye (Secale) only 1.4% of the bivalents are achiasmate. The remaining 98.6% of bivalents possess one (26.2%), two (70.5%), or three (1.7%) exchanges. These observations differ markedly from those predicted by a Poisson distribution of chiasmata per bivalent, in that there are large excesses of bivalents with one exchange and corresponding deficiencies of bivalents with zero or two exchanges. Similarly in Drosophila females, for the acrocentric X chromosomes in a genetically normal female, the frequency of achiasmate (E_0) bivalents is approximately 6–10%, the frequency of bivalents with a single exchange (E_1) is approximately 60–65%, and the frequency of bivalents with two exchanges (E_2) is approximately 30–35%. (Bivalents with three or more exchanges are observed only very rarely.) Each arm of the metacentric second and third chromosomes displays a similar pattern of exchange frequency, such that the E_0 frequency for each arm is approximately 10%. Thus, the probability of a major autosome being achiasmate is the product of the E_0 frequencies for each of the two arms, or approximately 2%.

The position of recombination events is also tightly controlled: exchange only occurs in the euchromatin, and the amount of exchange is not proportional to physical distance (Lindsley & Sandler 1977; Jones 1984; Berger et al. 2001). Rather, for each of the five major chromosome arms in Drosophila, the frequency of exchange is extremely low near the base and tip of the euchromatin and reaches its highest levels in an interval beginning approximately 30% of the distance from the tip to the base of the euchromatin, and ending some 50–60% of that distance (see figure 7.4). This pattern is not unique to Drosophila females, but rather is a general feature of chiasma distribution in a large number of organisms (Jones 1984).

The distribution of meiotic recombination results from the combined action of three types of genetic control: (i) *trans*-acting regulators of exchange position, which appear to act at the level of entire chromosome arms (Baker & Hall 1976); (ii) local *cis*-acting regulators of exchange [cf. Szauter (1984)]; and (iii) chromosomal elements, such as centromeres and telomeres, which suppress exchange in a polar fashion over long chromosomal distances. All three of these levels of regulation appear to act to keep exchanges, and hence chiasmata, a substantial distance away from the centromeres and the telomeres. Evidence in several organisms suggests that exchanges too close to the centromere may infer with homolog separation at meiosis I. Exchanges too close to the telomeres fail to properly orient homologous centromeres, perhaps as a consequence of insufficient sister chromatid cohesion distal to the chiasma (Koehler et al. 1996; Lamb et al. 1996; Orr-Weaver 1996; Ross et al. 1996a). The mechanisms by which these centromere and telomere effects are mediated remains obscure.

7.4.6 Practicalities of mapping

The effectiveness of any mapping method you use will only be as good as your markers, and your ability to score them. We can recall a failed, but instructive, attempt to map a mutant that affects meiotic chromosome segregation by using two flanking mutants that had identical phenotypic effects. The experiment's objective was to precisely map a meiotic mutant (*m*), which by itself produced no visible phenotype, by collecting recombinants between the genes *singed* and *forked*. As the meiotic mutant was known to lie between these two genes, it was hoped that by recovering a large number of recombinants between *singed* (*sn*) and *forked* (*f*) from + *m* + /*sn* + *f* mothers and testing whether or not each recombinant carried the mutant, one could position the meiotic mutant within this interval. Unfortunately, although the cross looked fine on paper, in reality it was impossible to distinguish the *singed–forked* recombinants from one of the parental chromosomes (*sn* + *f*), all three genotypes have a similar effect on bristle morphology. It was then necessary to restart the experiment, replacing the *sn* mutant with a closely linked mutant known as *cut* (*ct*) that affects wing morphology. The + *m* + /*ct* + *f* experiment worked nicely. But had the investigator thought beforehand about just how this cross was to be scored, they would have been spared much lost time and a fair amount of teasing. We tell this story to remind you that whatever markers you choose, you will need to score. *Accordingly, we urge you to stick to clean, easily observable markers.*

We also urge common sense in setting up the size of your experiment. Although methods exist for the statistical evaluation of recombination frequencies [cf. Szauter (1984)], it will be very hard to convince anyone that two map lengths of say 18 and 20 cm are meaningfully different, no matter how large your numbers are. Generally, most workers report map lengths to only figure to the right of the decimal point. Thus 1,000 progeny are fully sufficient. (Obviously, those looking for very rare exchange events will need larger numbers.). We also urge those who need to measure the length of long distances to use a large number of markers that lie within the larger interval.

7.5 The mechanism of recombination

7.5.1 Gene conversion

In this section, we will turn our attention from using recombination as a tool (i.e. mapping) to the mechanism of recombination itself. There are several recent reviews on the evolution of our current models of recombination [cf. Fogel et al. (1982); Szostak et al. (1983); Stahl (1994, 1996); Paques and Haber (1999)], and the interested reader is implored to read them. These are wonderful reviews that

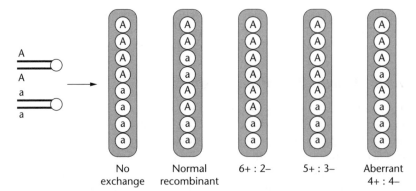

Figure 7.18   Gene conversion in Neurospora

include a depth of details and reveal the excitement of this process in a fashion that we can only hint at here.

The story of recombination models begins with the discovery by Lindegren of a phenomenon called **gene conversion**. This term described a case where sporulating an **Aa** diploid produced one of the three types of unusual asci diagrammed in figure 7.18. In the first case, known as a 6+:2− ascus, there are six **A** asci and only two **a**-bearing asci. [There is a reciprocal class (6−:2+) that is observed as well.] The easiest way to understand this result is to imagine that one of the **A** alleles had reached over and "*converted*" an **a** allele into an **A** allele. Hence the term gene conversion. That gene conversion was indeed precise was demonstrated in yeast (where the 6+:2− or 2+:6− segregations seen in Neurospora can be visualized as 3+:1− or 1+:3− asci) by Fogel and Mortimer (1969, 1970) using nonsense mutants. By demonstrating that the newly converted mutant spore was suppressible by the same tRNA suppressor that restored function to the original (converting) mutant, these authors were able to show that conversion was an accurate process in which information from one allele was faithfully replaced by the allele on its homolog.

As odd as the 6+:2− (or 6−:2+) segregations might be, the 5+:3− (or 5−:3+) and the aberrant 4+:4− segregations are even more curious. The only way to understand these observations is to suggest that at the end of meiosis one or more chromatids carried *both* the **A** and **a** alleles. The only way to understand this result is to suggest that these chromatids carried **A** information on one strand of the DNA duplex and **a** information on the other, such that replication of that chromatid produced both **A** and **a** haploid daughter cells. That is to say, the DNA of these chromosomes must include a region of base pair mismatch corresponding to the base pair(s) that differ between the **A** and **a** alleles. This structure is referred to as a **heteroduplex**. [Stahl (1969) has proposed an intriguing alternative (but sadly quite wrong) explanation for this observation that does not require heteroduplex formation.] The concept of heteroduplex creation and repair already had a long history in phage genetics (Wildenberg & Meselson 1975; Wagner & Meselson 1976).

It was clear from the beginning that gene conversion was associated with the process of recombination: most notably, gene conversion was accompanied by exchange of flanking marker in approximately 50% of the cases in which it occurred. Better yet, a substantial fraction of crossovers were associated with gene conversion events (Hurst et al. 1972; Borts & Haber 1987). In addition, those cases of gene conversion that were associated with flanking marker exchange displayed interference, while those not associated with interference did not. These observations led to the development of the first set of recombination models, which are described in the following section. (We might however note that far too much emphasis was placed on the observation that gene conversion was accompanied by exchange of flanking marker in approximately 50% of the cases in which it occurred. It was by no means true for all loci.)

7.5.2   *Previous models*

The first, and in many ways most important, model of recombination was proposed by Robin Holliday (Holliday 1964) and is referred to as the Holliday model. It is diagrammed in figure 7.19. The initiating step is an identical single-strand nick on both homologs. The two nicked single strands then swap Watson–Crick partners over a given distance of base pairs. In doing so they create two **symmetrical heteroduplexes**. Repair of both duplexes in the same direction (cf. A $\rightarrow$ a) could create a 6+:2− segregation event, repair of only one a 5+:3− event, and no repair might yield the aberrant 4+:4−.

Holliday proposed that the junction formed by strand exchange could be resolved by either of two types of DNA cutting–ligation events. Better yet, one of these created flanking marker exchange and the other did not. By imagining that the original double heteroduplex structure (known as a **Holliday intermediate**) was free to isomerize about the site of strand exchange (the **Holliday junction**), Holliday imagined that those two types of DNA cutting and ligation events might be equally frequent. This part of the model neatly explained the observations of Case and Giles (1958, 1959, 1964), Stadler and Towe (1963), and others that approximately 50% of irregular segregations were associated with reciprocal recombination between flanking markers.

Unfortunately, although molecules that could perform the necessary single-strand switching events were being isolated, it remained hard to understand how a pair of identical single-stranded nicks could be created. More troublesome for the Holliday model was a set of observations created by Rossignol and his collaborators regarding conversion at the b2 spore color gene in the fungus *Ascobolus immerses* (Rossignol et al. 1978, 1984). These authors observed a polarity of conversion frequencies in this gene, such that the frequency of conversion was far higher at one end of the locus than it was at the other. One could presumably account for this using the Holliday model by making the initial single-strand nicks site-specific. But Rossignol noted that the asymmetrical conversion events (5+:3− or 3+:5−) were clustered at the high end of the conversion gradient, while the symmetrical (aberrant 4+:4−) events were clustered at the low conversion end of the gradient (figure 7.20).

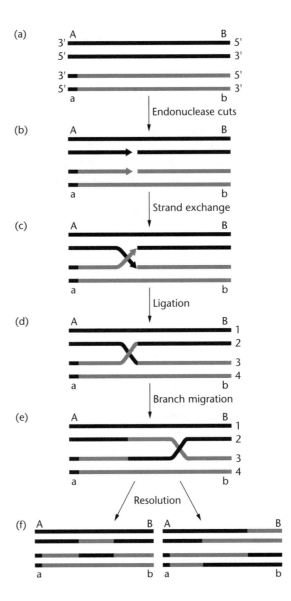

Figure 7.19   Holliday model

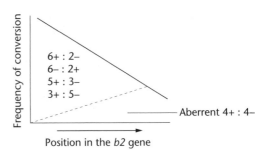

Figure 7.20   Polarity of
conversion in Ascobolus

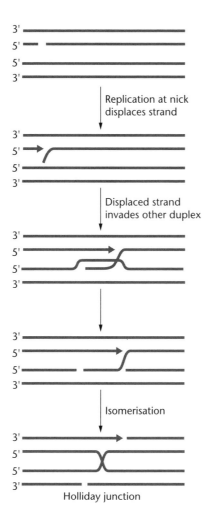

3'
5'
5'
3'

↓ Replication at nick
displaces strand

3'
5'
5'
3'

↓ Displaced strand
invades other duplex

3'
5'
5'
3'

↓

3'
5'
5'
3'

↓ Isomerisation

3'
5'
5'
3'

Holliday junction

Figure 7.21 The Meselson–Radding model. Recombination is initiated by a single-strand nick on only one of the two chromatids involved. (Please note the line pairs shown in this drawing are DNA duplexes, not sister chromatids.) Once the gap is created, DNA replication serves to displace the nicked strand 3′ to the nick. The freed single strand then invades the homologous duplex to create a D-loop structure. Nicking one of the two intact stands in the D-loop allows symmetrical strand displacement to create a Holliday intermediate.

Attempting to circumvent these difficulties, Meselson and Radding (1975) presented a truly clever model in which recombination was initiated by a single-strand nick on only one of the two homologs. This model is diagrammed in figure 7.21. The Meselson and Radding model differs primarily from its predecessor by producing primarily **asymmetric heteroduplex** regions (i.e. regions in which the heteroduplex is found on only one of the two chromatids involved in the recombination events). The absence of a significant amount of symmetrical heteroduplex neatly explained the absence of aberrant 4+:4– conversion events in yeast [see Fogel et al. (1982)]. By presuming that the asymmetrical heteroduplex that is produced is usually efficiently repaired in yeast, the preponderance of 6+:2– (or 6–:2+) events in yeast could be explained as well. The small amount of symmetrical heteroduplex produced near the Holliday junction itself seemed sufficient to explain the aberrant 4+:4– segregation events observed in other organisms.

Unfortunately, as attractive as the Meselson–Radding model was, it would fall prey to a series of observations that demonstrated the initiating event in meiosis was not a single-strand nick, but a double-strand break.

7.5.3   The currently accepted mechanism of recombination: the DSBR model

Several lines of evidence indicate that the initiating event in meiotic recombination is a double-strand break, not any type of single-strand nick. First, the directed gene conversion event that underlies mitotic mating-type switching in yeast was initiated by a double-strand break (Strathern et al. 1982; Klar et al. 1984). Second, the efficiency of integrative transformation in yeast was increased by a double-stranded break in the previously circular donor DNA, but not by a single-stranded nick (Orr-Weaver et al. 1981; Szostak et al. 1983). Third, meiotic recombination hot-spots were shown to correspond to hot-spots for double-strand breaks (not single-strand nicks) during early meiotic prophase (Nicolas et al. 1989; Sun et al. 1989; Cao et al. 1990).

All of these lines of evidence would lead to the presentation of the double-strand break repair (DSBR) (figure 7.22). That break is then resected to create a gap with two 3′ overhangs (White & Haber 1990). The invasion of those two ends into the intact homolog creates the **double Holliday** structure shown in the figure. This double Holliday structure has been isolated from meiotic cells by Schwacha and Kleckner (1994, 1995). [*An explanation for the manner in which the double-strand break model might produce the different kinds of conversion patterns seen in yeast and Ascobolus is provided by Nicolas and Petes (1994).*] Considerable evidence suggests that those gene conversion events not associated with crossover events may be created by a different pathway than those associated with flanking marker exchange (Allers & Lichten 2001). Specifically, non-crossover products appear during meiosis at the same time as the double Holliday junction intermediates, but crossover products are formed much later. To quote Allers and Lichten, "these results suggest that crossovers are formed by the resolution of Holliday Junction intermediates, while most noncrossover recombinants arise by a different, earlier pathway." Indeed, non-crossover events most likely occur by a process referred to as synthesis-dependent strand annealing (SDSA) (Belmaaza & Chartrand 1994; Nassif et al. 1994).

This view of separate origins for crossover and non-crossover events is consistent with the speculations of Carpenter (1987), that simple conversion events preceding events destined to facilitate crossover production are marked by a different type of recombination nodule, and perhaps occur by a different mechanism. Indeed, recent studies by Cromie and Leech (2000) on the biochemistry of recombination intermediate resolution provide important molecular clues regarding the mechanisms by which such intermediates are processed to produce either crossover or non-crossover products.

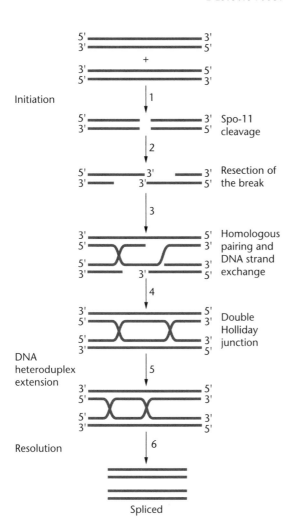

Initiation

Spo-11 cleavage

Resection of the break

Homologous pairing and DNA strand exchange

Double Holliday junction

DNA heteroduplex extension

Resolution

Spliced

Figure 7.22 The double-strand break repair model. Recombination is initiated by a double-strand break on only one of the two chromatids created by the Spo-11 enzyme. (Please note the line pairs shown in this drawing are DNA duplexes, not sister chromatids.) The break is then resected to form a gap with two 3′ overhangs. These two 3′ ends then invade the intact duplex to create symmetrical Holliday junctions.

Summary

This chapter has focused primarily on the use of recombination to generate maps and on the analysis of the recombination process itself. However, we remind the reader that the function of exchange is not to help us map genes, but to ensure chromosome segregation. We now move on in chapter 8 to discuss the mechanism by which that is accomplished.

8 Meiotic chromosome segregation

We have noted previously that the ability of chromosomes to divide reductionally at meiosis I is the physical basis of Mendelian inheritance. In this chapter we address the mechanism by which that separation takes place. The most critical issues to understand about meiotic chromosome segregation are the following:

1 *At the first meiotic division (MI) homologs separate to opposite poles of the spindle; MII is simply a haploid mitosis.* For the most part, the second meiotic division simply "processes" the complement of chromosomes delivered by the first meiotic spindle. However, chromosomes that have misbehaved at the first meiotic division are more likely to misbehave at the second meiotic division and in subsequent mitotic divisions.

2 *Proper segregation is mediated by forces (usually chiasmata) that act to hold bivalents together, and **not** by homology of centromeres.* Thus, two chromosomes that have undergone exchange will usually segregate from each other even if their centromeres are not homologous. Similarly, in the absence of a back-up system, non-exchange homologs will not properly segregate from each other even though their centromeres are homologous.

3 *The function of recombination is to ensure homolog segregation at anaphase I.* Thus, the failure of two homologs to undergo genetic exchange increases the likelihood that they will fail to segregate from each other properly at anaphase I (i.e. that they will **nondisjoin**). Recombination accomplishes this goal by using sister chromatid cohesion to create inter-homolog adhesion.

4 *Some organisms, such as Drosophila (Hawley & Theurkauf 1993; Hawley et al. 1993), yeast (Dawson et al. 1986; Guacci & Kaback 1991; Molnar et al. 2001), and C. elegans (Riddle et al. 1997), have back-up systems to ensure the segregation of those homolog pairs that might fail to recombine.* However, in all of those organisms, the back-up systems have a lower degree of fidelity than exchange-based (or chiasmate) segregation. Moreover, back-up systems clearly do not exist in at least some other organisms, such as higher plants, and in those organisms achiasmate chromosomes dissociate from each other prior to the first meiotic division and segregate at random with respect to each other at that division.

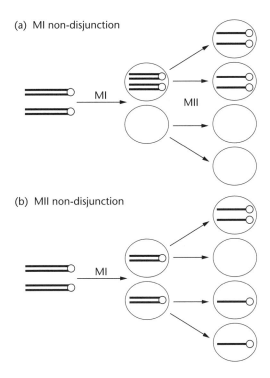

(a) MI non-disjunction

MI

MII

(b) MII non-disjunction

MI

Figure 8.1 Nondisjunction at meiosis I and meiosis II

8.1 Types and consequences of failed segregation

The failure of proper meiotic segregation is heralded by **nondisjunction**. We define nondisjunction as either the movement of both homologs to the same pole at meiosis I (MI nondisjunction), or the presence of two separated sister chromatids at the same pole at the end of anaphase II (MII nondisjunction; figure 8.1). We distinguish nondisjunction from chromosome loss events in which one homolog or one chromatid simply fails, for whatever reason, to reach a pole.

The result of nondisjunction at either division is a daughter cell carrying two copies of the nondisjoining chromosome (called a **diplo–** cell) or no copy of the chromosome in question (a **nullo–** cell). The results of the two nondisjunctional events differ in terms of the genetic composition of the diplo– products. In the case of MI nondisjunction, the two chromosomes will carry different centromeres; in the case of MII nondisjunction, both chromosomes will have come from the same homolog, and thus the two chromatids were once sisters. The preservation or loss of centromere heterozygosity thus becomes the defining tool for determining the division at which a diplo– exception arose (figure 8.2).

There is no method for determining absolutely when a nullo– exception arose. Indeed, nullo– exceptions can arise as simple chromosome loss, without nondisjunction. We distinguish between processes that result in nondisjunction and those that result in chromosome loss by looking at the ratio of diplo– to

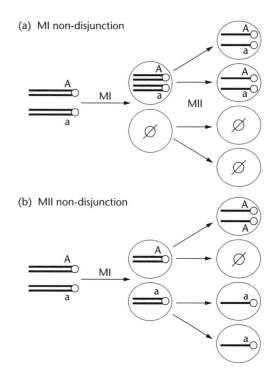

Figure 8.2 Centromere markers distinguish between nondisjunction at meiosis I and meiosis II

nullo– exceptions. In the absence of viability effects that favor one type of exceptional progeny or gamete over the other, nondisjunction will produce equal frequencies of nullo– and diplo– exceptions. Chromosome loss can produce only nullo– exceptions.

In general, meiosis I nondisjunction will be followed by a regular MII that produces two diplo– daughter cells and two nullo– daughter cells. In other words, the two chromosomes that nondisjoined at MI will normally behave properly at MII, each segregating one sister chromatid to each of the two daughter cells. There are cases where nondisjunction at MI predisposes a chromosome to loss or subsequent nondisjunction at MII (Hawley et al. 1993), but these situations are rare.

8.2   The origin of spontaneous nondisjunction

There have been several studies of spontaneous nondisjunction in Drosophila and humans, beginning with Bridges (1914, 1916). In Drosophila the most instructive of these is the report by Koehler et al. (1996). In humans the most comprehensive consideration is the studies of chromosome 21 nondisjunction by Lamb et al. (1996). The results of these two studies are described below.

MI exceptions: Koehler et al. (1996) demonstrated that the vast majority of spontaneous MI nondisjunctions in Drosophila oocytes (76.6%) result from the failure of the achiasmate back-up system (see below), which ensures the segregation of achiasmate homologs. The smaller proportion (23.4%) of spontaneous diplo-X exceptions derived from single exchange bivalents exhibit a strong bias toward very distal crossovers [see Table 2 and Figure 4a of Koehler et al. (1996)], which precisely parallels recent observations in human oocytes for chromosome 21 (Lamb et al. 1996) and chromosome 16 (Hassold et al. 1995). These data are also supported by findings in yeast suggesting that distal exchange bivalents are often susceptible to meiotic nondisjunction at MI (Ross et al. 1996a). Based strongly on the suggestion of Ross et al. (1996a,b), we will argue below that the cause of such nondisjunction is the lack of sufficient sister chromatid cohesion distal to the chiasma to ensure homolog adhesion to or through prometaphase, resulting in a bivalent that is effectively achiasmate. Thus, the failure of these segregations may also be thought of as a failure of the achiasmate system. This conclusion is consistent with the view that nondisjunction of bivalents with distal crossovers is far more common in Drosophila mutants that impair the achiasmate segregational system than it is in otherwise wildtype oocytes (Carpenter 1973; Rasooly et al. 1991). No such difference is seen for more medial or proximal crossovers.

The origin of MII exceptions: all of the MII exceptions obtained by Koehler et al. (1996) carry an exchange in the heterochromatin or very proximal euchromatin. A similar observation has been made by Lamb et al. (1996) for cases of MII nondisjunction of chromosome 21 in human oocytes. One explanation for both of these observations is the so-called "entanglement" model (Koehler et al. 1996; Lamb et al. 1996; Orr-Weaver 1996). According to this model, homologous chromosomes become entangled either as a consequence of multiple exchanges or, as we consider more likely, by an inability to resolve pericentromeric exchanges by anaphase I.

Chiasma resolution requires, at minimum, the release of sister chromatid cohesion distal to the site of exchange. Even if sister chromatid cohesion is released throughout the euchromatin at anaphase I, pericentromeric exchanges might not be resolved due to prohibition of sister chromatid cohesion release near the centromere. As a consequence, at MI the entire bivalent would either stall on the metaphase plate and be lost, or be dragged to a single pole, followed by disjunction of the homologous chromosomes at MII. The former event may explain the universally observed excess of nullo-X ova recovered over diplo-X ova in Drosophila. The latter event would effectively result in a delayed reductional division. Accompanied by sister chromatid splitting, usually the hallmark of MII, this phenomenon could cause two sister chromatids to end up in the same oocyte.

Koehler et al. (1996) and Lamb et al. (1996) also proposed an alternative model in which the resolution of pericentromeric exchanges prior to anaphase I results in precocious sister chromatid separation. According to this model, attempts to resolve extremely proximal chiasmata would frequently result in precocious sister chromatid separation within the heterochromatin before anaphase I, and subsequent dissociation of the two sister chromatids at the centromere of at least one homolog. If the sister chromatids separate precociously but continue on to the same pole at anaphase I, their random disjunction at anaphase II will result in *apparent* MII nondisjunction half of the time, as in the case of the six "MII" exceptions recovered in these experiments. If, on the other hand, the sister chromatids separate precociously and then go on to opposite poles at anaphase I, the result will be an exception heterozygous for proximal regions, similar to exceptions that have arisen through the more "classical" mechanism of nondisjunction at meiosis I. That nondisjunction can result from precocious sister chromatid separation has been demonstrated for two meiotic mutations, *ord* and *mei-S332*, whose products regulate the control of sister chromatid cohesion in Drosophila [cf. Lee and Orr-Weaver (2001)].

At present, it is not possible to distinguish between these two models. However, what both models have in common is a conflict between the need to resolve a proximal chiasma and the need to maintain sister chromatid cohesion in the pericentric heterochromatin. We propose that these putative difficulties inherent in the resolution of proximal chiasmata may represent a significant mechanism for meiotic nondisjunction. It is thus not surprising that in most organisms, meiotic recombination is highest per unit length in the medial regions of each pair of chromosome arms.

8.3   The centromere

8.3.1   *The isolation and analysis of the* S. cerevisiae *centromere*

Clarke and Carbon (1980) reported the first isolation of a centromere in yeast (and thus in any organism). Classical tetrad analysis had mapped the centromere of yeast chromosome III to the interval between the LEU2 and CDC10 genes. Clarke and Carbon created a "chromosome walk" of cloned DNA comprising all the sequences between these two genes, a 25-kb interval. The centromere had to have been in there somewhere.

The plasmid clones that comprise this walk are displayed in figure 8.3. Each of these clones carries a selectable marker (TRP1) that we can use to follow the gene

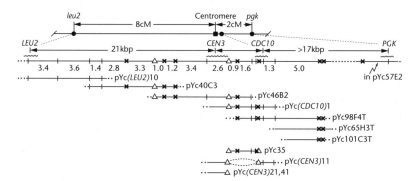

Figure 8.3   Genetic and physical maps of the centromere region of *S. cerevisiae* chromosome III studied by Clarke and Carbon. (After Clarke & Carbon 1980)

in yeast and a yeast DNA replicator sequence (*ars1*) that allows the plasmids to replicate in yeast. The clones also carried sequences required for propagation and selection in *E. coli*. Such plasmids were easily maintained in yeast, even in the absence of any kind of centromere insert, as long as one continued the selection in favor of the TRP1 gene or some other yeast gene carried by the plasmid. But as soon as one relaxed that selection (e.g. fed the cells Tryptophan), the plasmid was quickly lost in the yeast culture. (Why? The answer lies in the fact that these yeast divide by budding, and the daughter cell is much smaller than the mother cell. Daughter cells often inherit only a few or no copies of the autonomously replicating plasmid. If selection is relaxed, the plasmid-minus daughter cells will take over.)

It was exactly the weakness of these plasmids – their rapid loss in the absence of selection – that suggested a method of finding centromeres to Clarke and Carbon. Perhaps a plasmid carrying a functioning yeast centromere might function as a mini-chromosome and thus be stabilized by the mitotic machinery, even in the absence of selection. The test of that hypothesis was straightforward. Clearly the set of plasmid clones making up the 25-kb walk must contain the yeast centromere. Might one of those clones be stable in the absence of selection? Indeed, the pYe(CDC10)1 plasmid, which contains the CDC10 gene plus 8 kb of flanking yeast DNA, turns out to be special in exactly this sense. To quote Clarke and Carbon, "Unlike other *ars1* bearing plasmids . . . pYe(CDC10)1 is relatively stably maintained (in the absence of selection) through mitotic and meiotic cell divisions." By testing the mitotic stability of other plasmids with overlapping inserts, Clarke and Carbon were able to show that "the stabilizing segment is confined to a 1.6 kb fragment on the left (LEU2) side of CDC10."

But it is one thing to show that a given DNA sequence can confer mitotic stability on a plasmid, and another to claim that the sequence is a bona fide centromere. To make that claim the authors needed to prove that their 1.6-kb sequence had other properties of real centromeres, such as the ability to segregate properly at the two meiotic divisions. To test this hypothesis, Clarke and Carbon followed the segregation of a single copy of pYe(CDC10)1 in a diploid and followed the segregation of one of several other markers (e.g. *ade1* or *met14*) that define the centromeres of other chromosomes. Because this diploid has but one copy of pYe(CDC10)1, at meiosis I we expect the plasmid to go to only one pole. But at MII it ought to separate two sister chromatids and one should go to each pole. Thus following meiosis the spores should be 2+:2– for the pYe(CDC10)-born TRP1 marker. As shown in table 8.1, the plasmid segregated 2+:2– in 58–92% of the tetrads. The result is really better than that because we can ignore the cases of 0+:4– segregation. These diploids had probably lost the plasmid before the meiotic divisions started. Moreover, the "other centromere" data tells us that very few TTs were observed. That is the two plasmid-bearing cells in the 2+:2– tetrads were sisters from the last MII. To quote the authors, "Only parental ditype and nonparental ditype asci were obtained in this cross. . . . Thus the TRP1 locus on the plasmid behaves as a centromere linked marker."

The approach of Clarke and Carbon would eventually be used to map and sequence all 17 yeast centromeres. The minimal yeast centromere would be shown to be only 125 base pairs in length, and each centromere carried the same three functionally conserved domains (I, II, and III) [for a review, see Clarke and Carbon (1985); Hyman and Sorger (1995)]. Meiosis- and mitosis-specific domains have been identified within this centromere sequence, as have regions critical for sister chromatid cohesion. Moreover, the genetic and biochemical tractability of this simple centromere have allowed the identification of a number of genes whose protein products act at the centromere. One such study is detailed in box 8.1.

Unfortunately, the simplicity of the budding yeast centromere would not be a harbinger of things to come. The centromeres of higher eukaryotes would turn out to be rather more complex, and their understanding would require a major paradigm shift in modern biology. The story, described in the next section, begins in Drosophila.

Box 8.1   Identifying genes that encode centromere-binding proteins in yeast

Doheny et al. (1993) described two elegant assays to identify genes that affect kinetochore function in yeast: one assay detects relaxation of a transcriptional block that is observed at centromeres; the other detects an increase in the mitotic stability of dicentric chromosomes. Doheny and her collaborators used this assay to sort through a set of mutants recovered by Forrest Spencer on the basis of impaired transmission of mini-chromosomes (so-called *ctf* mutants). The first assay is based on screening for mutants that allow transcriptional read-through of a centromere, presumably by removing a centromere protein that serves to block polymerase progression. When transcription from a strong promoter is initiated towards a CEN DNA sequence, the mitotic segregational function of the centromere is destroyed. However, the chromatin structure of the kinetochore remains undisrupted, indicating that at least some of the centromere remains intact. Moreover, the majority of the transcripts terminate at the border of the CEN sequence. Thus the active polymerase is sufficient to kill centromere activity, but the remaining centromere structure is sufficient to impede the polymerase.

Doheny et al. choose to exploit this stand-off by screening for mutants that allow read-through across such an inactive centromere construct that

has been integrated into one of the arms of a yeast chromosome. (Because the centromere is inactive, we can integrate such constructs into a normal yeast chromosome without creating a dicentric.) One such construct is shown in figure B8.1. The promoter carried by this construct (GAL) is inducible and very strong. Moreover, lying beyond the CEN sequence is a transcription mutant capable of changing the color of the yeast colony: read-through equals color change. But inserting the 165 bp of CEN6 knocks down transcription by 100×, blocking that color change. To show that the transcriptional block is really due to at least partial kinetochore assembly, Doheny replaced the normal CEN sequence with a single mutant in CDEIII. This mutant reduces mini-chromosome transmission by 220× so it clearly impairs kinetochore function. The use of this mutant insertion elevated levels of transcriptional read-through to 20% of normal. This change was sufficient to create a color change in colonies, providing a rapid tool for screening through the *ctf* mutants – just insert the construct carrying the normal CEN6 construct into yeast carrying each of the *ctf* mutants. This test identified seven possible candidate mutants from among the available collection of *ctf* mutants.

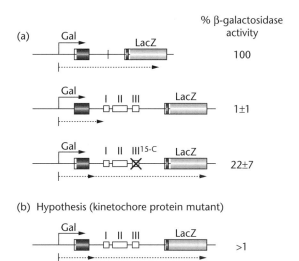

Figure B8.1 Transcriptional read-through reporter construct and assay. (After Doheny et al. 1993)

This screen tested only for mutants in genes whose products bound to the CEN sequence. The second was a lovely, if counterintuitive, screen for those mutants affecting centromere function. The basic idea is that a mutant that impairs centromere function ought to restabilize a dicentric chromosome by making one of the two centromeres "weaker." To perform this experiment, Doheny inserted her regulatable centromere construct into a dispensable chromosome fragment (a partial disome) that carries both a functional CEN sequence and a useful reported gene. She could then activate that centromere by repressing the activity of the GAL promoter. The result was a dicentric chromosome that was quickly lost. But mutants that disrupted CEN function weakened the expression of one or both of those centromeres, and stabilized the chromosome. Indeed, these two assays together identified two genes of interest, at least one of which is now known as a component of the centromere.

8.3.2 The isolation and analysis of the Drosophila centromere

Gary Karpen began the assault on the fly centromere using the smallest known stable chromosome, *Dp(1;f)1187* (Murphy & Karpen 1995; Le et al. 1995; Sun et al. 1997). As shown in figure 8.4, this small chromosome was derived from a normal sequence X chromosome by an X-ray-induced inversion and a subsequent X-ray-induced deletion. Although it is quite small, the *Dp(1;f)1187* chromosome can easily be followed in crosses because it carries a useful visible marker (*y+*) at its left tip. Although this chromosome is only approximately 1 Mb in length, it segregates faithfully at both meiosis and mitosis. In order to be able to create deletion derivatives of this chromosome, Karpen created a variant, known as *Dp8-23*, that carries two marked (*rosy+*) transposon insertions distal to the *y+* marker. (One needed at least a second scorable marker here in order to screen for further deletions of this chromosome.) Karpen now screened for both

Table 8.1   Meiotic segregation of the mini-chromosomes. [After Table 2 in Clarke and Carbon (1990)]

Genetic cross no.	Mini-chromosome in cross	Mini-chromosome marker scored	Distribution in tetrads of genetic marker on mini-chromosome (%)					Test for centromere linkage of marker on mini-chromosome		
			4+:0–	3+:1–	2+:2–	1+:3–	0+:4–	PD	NPD	T
1	pYe(CDC10)1	TRP1	1 (6%)	0	10 (63%)	0	5 (31%)	2	8	0
2	pYe(CDC10)1	CDC10	1 (8%)	0	11 (92%)	0	0	2	8	1
3	pYe(CDC10)1	TRP1	1 (7%)	0	11 (79%)	0	2 (14%)	4	7	0
4	pYe(CDC10)1	TRP1	4 (21%)	0	11 (58%)	0	4 (21%)	ND	ND	ND
5	pYe(CEN3)11	TRP1	2 (13%)	0	9 (60%)	0	4 (27%)	4	5	0

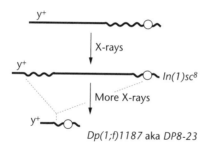

Figure 8.4   The origin of *Dp(1;f)1187*

terminal (loss of the *rosy*+ marker) and internal deficiency derivatives (loss of the *y*+ markers) on this chromosome.

In order to further define the centromere sequence, Karpen created a set of gamma-ray-induced deletion derivatives of the *Dp8-23* chromosome (Murphy & Karpen 1995). He then tested the stability of these derivatives by testing their ability to be properly transmitted through meiosis in males or females when present as a single copy – the monosome transmission test. (Remember that to even be recovered these derivatives had to be at least mitotically stable.) The monosome transmission test looks at the fraction of progeny that receive the derivative from a male or female parent carrying a single copy of that chromosome. A fully functioning chromosome would yield a value close to 50%. These authors also characterized the structure of these deletion derivatives by pulse-field electrophoresis.

Surprisingly, as shown in figure 8.5, deletions removing nearly half of the DNA content of this chromosome were fully meiotically and mitotically stable. To quote Murphy and Karpen (1995), "The normal behavior of all the *Dp(1;f)1187* derivatives . . . , including the 620 kb (deletion) derivative gamma-1230, indicates that the entire euchromatin, sub-telomeric heterochromatin and telomere, and over one-half of the *Dp(1;f)1187* centromeric heterochromatin are dispensable for chromosome stability." Thus sequences necessary for centromere functioning must reside in the 420-kb region. But the real find obtained by creating this set of deletion derivatives was not a deletion at all. Rather it was an inversion, referred to as gamma-238, in which the *y*+ and *rosy*+ markers marked separate ends of the chromosome. Now that both ends were marked, a second screen for gamma-ray-induced deletions would allow Karpen and his collaborators to obtain a series of deficiencies extending into the chromosome from both ends. The structure and transmissibility of those derivatives is described in figure 8.6 [Figure 4 of Murphy and Karpen (1995)].

A look at figure 8.6 suggests that full stability of a deletion derivative does require a small region of the original *Dp(1;f)1187* chromosome that Murphy and Karpen refer to as the Bora Bora island. All derivatives (Group A and Group B) that retain that region are fully stable regardless of what else they have lost. Moreover, deficiencies that have lost some fraction of the Bora Bora interval (Group C) show reduced level of transmission, especially in females. All of this would lead Murphy and Karpen to conclude that "completely normal chromosome stability requires Bora Bora plus flanking heterochromatin."

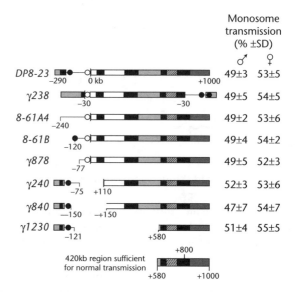

Figure 8.5 Karpen's first set of deletions. (After Murphy & Karpen 1995)

That argument was fine, and not the least heretical. At the time the Murphy and Karpen paper was published, the scientific community was still wrapped in a view that could be summarized as "in yeast, as in all creatures great and small." Flies had to have a yeast-like discrete centromere, it just might be harder to find. The problem with that argument was the recovery of the Group D derivatives. These elements are at least weakly transmissible; after all Karpen could recover them, keep them in stock, and test them. But they do not contain the Bora Bora island. Indeed, some of them seem to contain only material that was telomeric on the original *Dp(1;f)1187* chromosome and which could be deleted without consequence for transmission. (Compare the top two Group A derivates with the bottom five Group D derivatives.) Karpen explained these outliers by suggesting that they corresponded to transmissible acentric fragments, which might be maintained by some other means. This is *not* to say that they don't have some centromere activity, but rather just to say that they do *not* contain the normal centromere.

Yet a subsequent paper (Williams et al. 1998) would show that these odd acentric fragments possessed many of the properties expected of centromere-bearing chromosomes. These presumably acentromeric fragments bound the centromere-specific protein ZW10 and associated with the spindle poles at anaphase. Because these derivatives contain DNA normally found near the tip of the X chromosome, the authors suggest that these sequences have *acquired* centromere function. This is important because Williams et al. (1998) had shown that these fragment chromosomes carry only sequences that are usually found near the X chromosome telomere. Such fragments of telomeric DNA do not bind to ZW10, and if they are simply "broken-off" the end of the X, they only acquire this ability if they were derived from a region near the centromere. Indeed, Maggert and Karpen (2001) would show elegantly that a usually

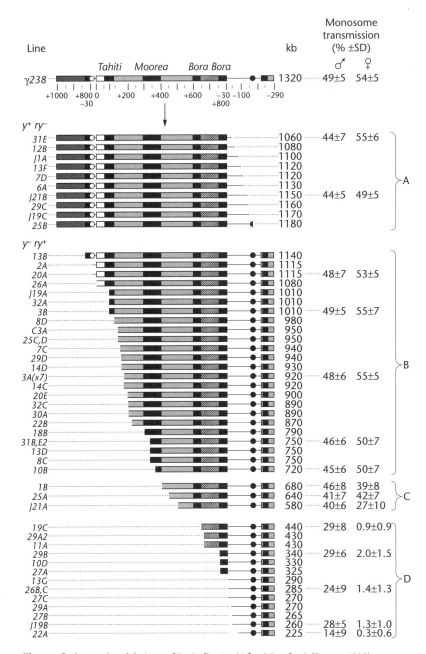

Figure 8.6 Further deletions of *Dp(1;f)1187*. (After Murphy & Karpen 1995)

non-centromeric sequence needed to be in close proximity to a centromere sequence in order to *acquire* centromeric activity (box 8.2).

These findings have led to a critically important model of centromere structure and function, in which the centromere is a state of chromatin rather than a specific sequence (Karpen & Allshire 1997). In this model the centromeric states become stable and heritable by virtue of the duplication of specific chromatin structures during each cell cycle. Rarely can sequences that do not normally function as centromeres acquire centromere function if they lie close to functioning centromeric intervals. Centromere function can also presumably be lost at low frequency, but this is usually a lethal event. Further data in support of this model, and its implications, are described in box 8.2.

Box 8.2   The concept of the epigenetic centromere in Drosophila and humans

The finding by Karpen and his collaborators that sequences born by the Group D derivative chromosomes shown in figure 8.6 might have *acquired* centromere activity was stunning. The only reasonable interpretation of these data was that being "a centromere" in flies was less about possessing a specific sequence than it was about possessing a certain "state." The idea was as heretical as concepts come, and yet it was strongly supported by four types of evidence:

1   The acentric chromosomes are transmitted at high frequency in several types of cell divisions. However, they are quite unstable in other types of divisions. For example, they are transmitted moderately well in male meiosis but fare quite poorly in female meiosis. They seem to do well in pre-blastoderm mitoses, but are frequently lost in neuroblast meiosis.

2   These acentric chromosomes bind a centromeric marker protein (ZW10).

3   Female transmission is greatly reduced in *nod/+* heterozygotes, demonstrating that segregation is mediated by movement on the spindle. Similarly, transmission is enhanced in females carrying extra doses of *nod+*. (The *nod* gene produces a chromosomal protein required to hold chromosomes on the developing spindle by opposing the poleward forces exerted by kinetochores, see below.)

4   Molecular analysis demonstrates that these chromosomes have not acquired new centromeric sequences.

To explain these observations, Karpen suggested that these chromosomes are a reflection of "neocentromeric activity," and that centromeric activity is not itself the direct property of some particular DNA sequence. Rather, centromeric activity reflects an epigenetic state that is stably passed on with each cell division. He further imagined, and later demonstrated (Maggert & Karpen 2001) that centromeric activity can spread to nearby sequences. The concept of an "*epigenetic centromere*" is fully presented in a seminal review by Karpen and Allshire (1997). These authors noted that viewing the centromere as a state, instead of a sequence, allowed one to understand rather a lot of previously confusing observations regarding the centromeres of other higher eukaryotes. For example, the centromeres of *S. pombe* were much larger and more complex than their *S. cerevisiae* counterparts (Clarke et al. 1993; Clarke 1998). Moreover, when centromere repeat-bearing sequences are transformed back into *S. pombe* cells they acquire centromere competence only rarely, but without sequence change (Steiner & Clarke 1994). Similarly, the analysis of small human marker chromosomes suggested that such chromosomes can acquire centromere function without acquiring the alpha-satellite DNA sequences present at normal centromeres (Karpen & Allshire 1997).

The model proposed by Karpen and Allshire (1997) is simple: centromeric activity in higher eukaryotes reflects a state of chromatin, not just a DNA sequence. That state is self-perpetuating at the time of DNA replication. This is not to say that the content of certain regions might favor the assembly or maintenance of that chromatin state, but only that such sequences are not required.

8.4   Segregational mechanisms

The world would be a much simpler and perhaps happier place if there were but one method of ensuring proper homolog separation at meiosis I.[1] Sadly, there are a large number of ways that evolution has created to accomplish this single task. None the less, in the vast majority of meiosis in species that we care most about, segregation is mediated by chiasma function. For that reason, we shall divide the universe of segregational mechanisms into two camps: chiasmate segregation and achiasmate segregation. Both types are reviewed briefly below.

8.4.1   *How chiasmata ensure segregation*

In the previous chapter we cited the work of Bruce Nicklas (1974, 1977) as demonstrating that exchanges ensure homolog separation at the first meiotic division. Because the sister chromatids display tight cohesion during meiotic prophase and metaphase I, chiasmata hold the paired chromosomes together and thus commit the two centromeres physically linked by the exchange to orient toward opposite poles of the spindle. However, the exact mechanism by which this is achieved does differ among meiotic systems. In plants and most animal male meioses, the spindle is assembled by centrosomes. In such meiosis, each kinetochore captures (or is captured by) microtubule fibers emanating from the kinetochore immediately after nuclear envelope break-down. Thus, each centromere of the bivalent attaches to the spindle and begins to move toward one of the two poles. Most bivalents achieve a bipolar orientation immediately, because at the start of prometaphase the two homo-logous centromeres are usually oriented in opposite directions, such that if one centromere is pointed at one pole the other centromere is pointed at the opposite pole (Nicklas, 1967). Once the two centromeres have oriented to-ward opposite poles, the progression of the two centromeres toward the poles is halted at the metaphase plate by the chiasma. This represents a stable posi-tion in which the bivalent will remain until anaphase I. If the bivalent does not immediately acquire a bipolar orientation, centromeres are capable of breaking and reforming spindle attachments until a stable position at the metaphase plate is achieved (Nicklas 1967; Nicklas & Staehly 1967; Nicklas & Koch 1969).

   In the meiosis of many females, meiosis is acentriolar. The chromosomes themselves organize the spindle. Initially each pair of homologs establish two oppositely oriented half spindles that then coalesce into a bipolar spindle [cf. Theurkauf and Hawley (1992); Hunt and LeMaire-Adkins (1998)]. Perhaps because it is opposed centromeres that facilitate bipolar assembly in female meiosis, observations of initial malorientation are rare.

---

[1]   Or of regulating gene expression for that matter.

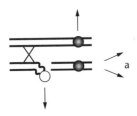

Figure 8.7 Exchange in a translocation heterozygote commits non-homologous centromeres to segregate from each other

The ability of chiasmata to link centromeres together does *not* require that the centromeres themselves be homologous. Homologous exchanges that conjoin (link) heterologous centromeres together will ensure that those chromosomes segregate from each other. As shown in figure 8.7, the exchange between two homologous euchromatic regions of an *X-4* translocation in Drosophila will commit the X and 4th centromeres to segregate from each other quite faithfully (Parker & Williamson 1976). Similarly, centromere "replacement" studies in yeast (Clarke & Carbon 1985) have shown that crossover bivalents segregate faithfully even when their centromeres are originally derived from different chromosomes.

And if chiasmata do not form? Although some organisms possess segregational back-up systems to ensure the segregation of achiasmate chromosomes (Hawley et al. 1993; Wolf 1994), many do not. In such organisms achiasmate bivalents fall apart at diplotene diakinesis, resulting in the premature dissociation of the bivalent into two univalents prior to prometaphase I. The two prematurely separated homologs align separately on the spindle, and stability at the metaphase plate is never achieved. Rather, both chromosomes move up and down the spindle, frequently reorienting. Their disjunctional fate at anaphase I is determined solely by their orientation on the spindle at the time anaphase begins. In this instance normal disjunctional events and nondisjunctional events (in which both homologs proceed to the same pole) will occur with equal frequency. Workers in a variety of organisms have also shown that such univalents are frequently unstable; such that chromosome loss, and breakage, are not uncommon events at both meiotic divisions and in subsequent mitotic divisions [for a review, see Hawley et al. (1993)]. Thus, chiasmata provide an essential device for stabilizing bivalents at the metaphase plate during prometaphase.

8.4.2 *Achiasmate segregation*

As noted above, many organisms have elaborate mechanisms to ensure the segregation of achiasmate chromosomes. Here we describe only those mechanisms that have been described in the most commonly utilized model organisms.

Achiasmate segregation in Drosophila males

There is no recombination in wildtype *D. melanogaster* males.[2] How then do these males carry out meiotic chromosome segregation? Unfortunately we cannot offer a general answer to this question, but rather must answer it on a chromosome-by-chromosome basis.

Two chromosomal sites, known as **collochores**, mediate sex chromosome segregation in Drosophila males (Cooper 1964). (The word collochore is Latin

---

[2] And we mean none. Nor is a synaptonemal complex produced.

for sticky spot.) The collochores map to the tandemly repeated rDNA present in the X heterochromatin and on the short arm of the Y chromosomes (McKee & Karpen 1990; Park & Yamamoto 1995) is in fact comprised of many copies of a 240-bp intergenic spacer sequence within the rDNA repeats found on both the X and Y chromosomes (McKee et al. 1992; Ren et al. 1997; McKee 1998). In this curious case, the remnants of a once common nucleolus seem sufficient to function as a pseudo-chiasma and function to hold the homologs together until they separate at anaphase I.

The mechanism by which the achiasmate autosomes are segregated remains even more mysterious. These chromosomes clearly pair along their lengths at metaphase. A cytogenetic study of the effects of duplications and deletions on this process has led McKee and his collaborators to conclude that it is based on widespread homology, and not on the function of specific pairing sites (McKee et al. 1993). At least one mutant, known formerly as *mei-S8* and now as *teflon*, specifically impairs autosomal pairing in Drosophila males (Sandler et al. 1968; Tomkiel et al. 2001). [Note added in proof: Since the submission of this manuscript, an excellent paper describing the live analysis of meiotic pairing in Drosophila males has been published by Vazquez et al. (2002).]

Achiasmate homologous segregation in Drosophila females

Despite the fact that exchange is fully sufficient to ensure segregation in Drosophila females, it is none the less clear that exchange is not always required to ensure the segregation of homologous chromosomes. For example, the obligately achiasmate 4th chromosome segregates normally from its homolog in greater than 99.9% of meioses. Similarly, non-exchange X chromosomes segregate from each other quite properly, even in the presence of heterozygosity for inversions (Sturtevant & Beadle 1936; Cooper 1964). Three lines of evidence suggest strongly that these achiasmate segregations are mediated by heterochromatic pairings. The first evidence that heterochromatic homology plays a crucial role in achiasmate segregations came from the analysis of the segregation of two normal 4th chromosomes in the presence of a homologous duplication (figure 8.8). These six duplications, known as $Dp(1;4)$s, all carry some or all of the heterochromatic base of the 4th chromosome capped by X chromosome euchromatin. Each was tested for its ability to ensure 4th chromosome nondisjunction as a result of $Dp(1;4) \leftrightarrow 44$ segregation events. Basically, the more 4th chromosome heterochromatin carried by a given $Dp(1;4)$, the higher the observed level of induced 4th chromosome non-disjunction (Hawley et al. 1992).

Karpen et al. (1996) presented stronger evidence that achiasmate segregation is dependent on heterochromatic homology. They showed that the frequency with which two achiasmate deletion derivatives of $Dp(1;f)1187$ (see above) segregate from each other in female meiosis is proportional to the amount of centric heterochromatic homology. Normal segregation requires 800 kb of overlap in the heterochromatin surrounding the centromere, whereas nearly

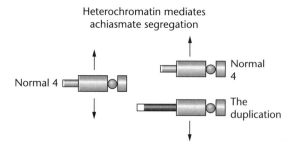

Heterochromatin mediates
achiasmate segregation

Normal 4

Normal
4

The
duplication

Figure 8.8 A duplication carrying 4th chromosomal heterochromatin can compete with a normal 4th chromosome for its partner

random disjunction is observed with only 300 kb of overlap. A linear correlation between the amount of heterochromatic homology and segregation efficiency was observed in the range of 800 to 300 kb. They concluded that sequences found throughout the centric heterochromatin of *Dp(1;f)1187* act additively to ensure achiasmate meiotic segregation. Karpen's work here is critical for two reasons. First, it provides the strongest possible evidence that centromere homology alone is not sufficient to guarantee segregation. Second, it clearly demonstrates the direct role of heterochromatic homology in ensuring achiasmate segregation.

Although heterochromatic pairings were known to exist as late as the end of pachytene (Carpenter 1975, 1979a,b), it was not clear whether or not they might persist during metaphase, the time at which segregational orientation is presumably determined and critical. Dernburg et al. (1996) used three-dimensional fluorescent hybridization to investigate the physical associations between achiasmate homologs from the end of pachytene until the onset of achiasmate segregation at prometaphase. Although euchromatic pairings dissolve following pachytene, heterochromatic pairings are preserved within the karyosome until prometaphase. Thus, in Drosophila meiosis, heterochromatic pairings persist beyond the dissolution of the synaptonemal complex at pachytene until centromere co-orientation at prometaphase.

In most organisms, paired but achiasmate bivalents would precociously dissociate as a consequence of homolog–homolog repulsion at diplotene diakinesis. However, in Drosophila female meiosis there is no stage comparable to diplotene diakinesis. Instead, after pachytene the chromosomes condense into the karyosome, where they remain until nuclear envelope breakdown and spindle formation during prometaphase. Karyosome formation occurs in stage 3 oocytes and the karyosome persists until the beginning of prometaphase in stage 13 oocytes. It seems likely then, as a consequence of this curious meiotic detour into the karyosome, that the pairings that existed at the end of pachytene, and most especially heterochromatic pairings, may persist until the beginning of prometaphase. This would allow paired, but achiasmate, bivalents to maintain centromere apposition until spindle assembly at prometaphase.

Drosophila also possesses a second, very different, achiasmate segregation system that allows the faithful segregation of two non-recombining, non-homologous chromosomes. That system is discussed in box 8.3.

Box 8.3   Achiasmate heterologous segregation in Drosophila females

The curious pattern by which Drosophila oocytes build and organize their spindle may also explain perhaps the most vexing of all meiotic events in Drosophila, namely the ability of the oocyte to segregate any two non-homologous chromosomes that would otherwise lack a partner (Grell 1976). For example, a compound X chromosome will virtually always segregate from a compound 4th chromosome. Indeed, two compound autosomes will virtually always segregate from each other regardless of their identity. The participation of such compound chromosomes in achiasmate disjunctions tells us two important things. First, the ability of any two compound chromosomes to segregate from each other, regardless of their genetic content, demonstrates that this process does not require homology. Second, the fact that even compounds which have undergone inter-arm exchanges participate in achiasmate segregations differentiates between the requirement for recombination *per se* and the requirement for chiasmata; only the latter are sufficient to prevent a compound chromosome from undergoing heterologous segregations. Like achiasmate homologous segregations, these non-homologous events require the function of the NOD protein. However, unlike achiasmate homologous segregations, these segregations are not preceded by physical interactions between the partner chromosomes either in pachytene or during prometaphase (Dernburg et al. 1996).

Hawley and Theurkauf (1993) have presented a model for heterologous segregation in which heterologous chromosomes remain randomly oriented until the spindle is assembled, and then orient such that a given chromosome orients toward the least crowded pole. This model is based on two well-documented properties of the meiotic spindle in Drosophila females. First, the movement toward the pole is size-dependent in that smaller chromosomes leave the metaphase plate earlier and move farther toward the pole than do larger achiasmate chromosomes (Theurkauf & Hawley 1992). According to this model a given non-exchange chromosome will move to the least crowded pole. Given the narrowness of the meiotic spindle in Drosophila females relative to the width of a chromosome, the physical basis of this blockage or retardation by other chromosomes might be nothing more than a physical obstruction of the ability of kinetochore microtubules to reach the poles. Accordingly, when there are only two non-exchange chromosomes, they always segregate from each other. That this model may indeed be correct is suggested by the live studies of achiasmate segregation in *S. pombe* (Molnar et al. 2001).

Achiasmate segregation in *S. cerevisiae*

Dawson et al. (1986) provided the first report of achiasmate segregation in budding yeast. Using circular mini-chromosomes with limited homology they were able to observe that the two chromosomes segregated faithfully, without benefit of chiasmata, in 100 out of 112 cases. The remaining 12 meioses included eight cases of meiosis I non-disjunction and four cases of non-disjunction at meiosis. They also demonstrated that this process was independent of homology; that is to say, in a diploid with three mini-chromosomes, only two of which were homologous, the two homologs segregated from each other no more than 33% of the time. Guacci and Kaback (1991) went on to show that "authentic" full-length yeast chromosomes could be segregated by this method as well, and confirmed that neither size nor homology were critical to ensure segregation [see also Kaback (1989) and Ross et al. (1996b)]. Curiously enough, however, these non-homologous segregations are preceded by non-homologous pairings. That is to say, in diploid monosomic from chromosomes I and III, the eventual segregation of chromosome I from chromosome III in almost 90% of the meioses is preceded by the pairing or tight association of these chromosomes at pachytene (Loidl et al. 1994).

Achiasmate segregation in *S. pombe*

Several studies of recombination-deficient mutants in *S. pombe* have revealed the existence of an achiasmate segregation system that may well be homology-dependent. The evidence for this assertion is that achiasmate homologs appear to segregate from each other faithfully in rec– cells in more than 75% of the meioses (Davis & Smith 2001). This is true despite the fact that two small mini-chromosomes do not segregate faithfully in this organism (Niwa et al. 1989). Achiasmate segregation can indeed be observed cytologically in this organism, and appears to function by a crowded pole-like mechanism (Molnar et al. 2001).

Summary

In this chapter we have given you a picture of the mechanisms that ensure segregation. It is segregation that explains Mendelian assortment, and that brings us back to the beginning. Perhaps this is a good place to end.

Epilogue

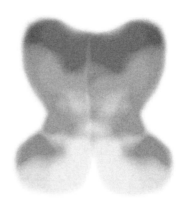

The preceding eight chapters cover the basic intellectual operations that comprise modern genetics. Some of these techniques, such as mutant hunting, suppressor analysis, and complementation analysis, echo issues derived from even the most current journals. Meanwhile others, such as the algebraic analysis of recombination data, have fallen into disuse. While writing this book, we were often reminded by our colleagues that some things are no longer done in the fashions described. This is after all the post-genomic era.

Living in an era of sequenced genomes is a heady business, indeed. But we are reminded of a comment by the playwright Noel Coward. When pressured by a friend (probably a publisher at Blackwell) about how he was doing on his latest play, he is reported to have answered that he was half-done. He had taken all the words out of the dictionary, now he just needed to put them in a proper order. The sequencing of many genomes has given us our list of words or genes. Now we, like Mr. Coward, need to put our words in the right order. Unfortunately, unlike Mr. Coward, we have been given many words that we do not understand. We must learn what these words (or genes) mean or do. We suspect that the best way to understand what genes do is to mutate them. And the best way to put them in order will be by epistasis analysis. One may identify interacting proteins by systems such as the yeast two-hybrid assay or mass spectroscopy, but those interactions will have real meaning only when confirmed by genetic interactions as well.

In that sense then, perhaps the genetics "that was" becomes the genetics "that is." We suspect that these tools, and the intellectual principles that created them, will have a much more than historical value for generations of biologists to come.

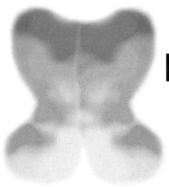

References

Adams, A. E. and Botstein, D. (1989). Dominant suppressors of yeast actin mutations that are reciprocally suppressed. *Genetics* 121: 675–83.

Adato, A., Kalinski, H., Weil, D., Chaib, H., Korostishevsky, M., and Bonne-Tamir, B. (1999). Possible interaction between USH1B and USH3 gene products as implied by apparent digenic deafness inheritance. *Am J Hum Genet* 65: 261–5.

Albertson, D. G. and Thomson, J. N. (1993). Segregation of holocentric chromosomes at meiosis in the nematode, *Caenorhabditis elegans*. *Chromosome Res* 1: 15–26.

Albertson, D. G., Rose, A. M., and Villeneuve, A. M. (1997). Chromosome organization, mitosis, and meiosis. In: D. L. Riddle, T. Blumenthal, B. J. Meyer, and J. R. Priess (eds.), *C. Elegans II*. Cold Spring Harbor Press, New York.

Allers, T. and Lichten, M. (2001). Differential timing and control of noncrossover and crossover recombination during meiosis. *Cell* 106: 47–57.

Anderson, E. G. (1925). Crossing over in a case of attached X chromosomes in *Drosophila melanogaster*. *Genetics* 10: 403–17.

Appling, D. R. (1999). Genetic approaches to the study of protein–protein interactions. *Methods* 19: 338–49.

Armstrong, S. J., Franklin, F. C., and Jones, G. H. (2001). Nucleolus-associated telomere clustering and pairing precede meiotic chromosome synapsis in *Arabidopsis thaliana*. *J Cell Sci* 114: 4207–17.

Arrizabalaga, G. and Lehmann, R. (1999). A selective screen reveals discrete functional domains in Drosophila Nanos. *Genetics* 153: 1825–38.

Ashburner, M. (1989). *Drosophila. A Laboratory Handbook.* Cold Spring Harbor Laboratory Press, New York.

Avery, L. and Wasserman, S. (1992). Ordering gene function: the interpretation of epistasis in regulatory hierarchies. *Trends Genet* 8: 312–16.

Avery, O., MacLeod, C. M., and McCarty, M. (1944). Studies on the chemical nature of the substance inducing transformation of pneumococcal types. *J Exp Med* 79: 137–58.

Babu, P. and Bhat, S. (1980). Effect of zeste on white complementation. *Basic Life Sci* 16: 35–40.

Baker, B. S. (1975). *Paternal loss (pal)*: a meiotic mutant in *Drosophila melanogaster* causing loss of paternal chromosomes. *Genetics* 80: 267–96.

Baker, B. S. and Carpenter, A. T. (1972). Genetic analysis of sex chromosomal meiotic mutants in *Drosophilia melanogaster*. *Genetics* 71: 255–86.

Baker, B. S. and Hall, J. C. (1976). Meiotic mutants: genetic control of meiotic recombination and chromosome segregation. In: M. Ashburner and E. Novitski (eds.), *The Genetics and Biology of Drosophila*. Vol. 1a. Academic Press, London, pp. 352–434.

Baker, B. S. and Ridge, K. A. (1980). Sex and the single cell. I. On the action of major loci affecting sex determination in *Drosophila melanogaster*. *Genetics* 94: 383–423.

Barlow, A. L. and Hulten, M. A. (1998). Crossing over analysis at pachytene in man. *Eur J Hum Genet* 6: 350–58.

Barton, N. H. and Charlesworth, B. (1998). Why sex and recombination? *Science* 281: 1986–90.

Baudat, F., Manova, K., Yuen, J. P., Jasin, M., and Keeney, S. (2000). Chromosome synapsis defects and sexually dimorphic meiotic progression in mice lacking Spo11. *Mol Cell* 6: 989–98.

Baumann, P. and Cech, T. R. (2001). Pot1, the putative telomere end-binding protein in fission yeast and humans. *Science* 292: 1171–5.

Beadle, G. W. and Emerson, S. (1935). Further studies of crossing over in attached-X chromosomes of *Drosophila melanogaster*. *Genetics* 20: 192–206.

Beall, E. L. and Rio, D. C. (1996). *Drosophila* IRBP/Ku p70 corresponds to the mutagen-sensitive mus309 gene and is involved in P-element excision in vivo. *Genes Dev* 10: 921–33.

Belmaaza, A. and Chartrand, P. (1994). One-sided invasion events in homologous recombination at double-strand breaks. *Mutat Res* 314: 199–208.

Belote, J. M., McKeown, M. B., Andrew, D. J., Scott, T. N., Wolfner, M. F., and Baker, B. S. (1985). Control of sexual differentiation in *Drosophila melanogaster*. *Cold Spring Harbor Symp Quant Biol* 50: 605–14.

Bender, A. and Pringle, J. R. (1991). Use of a screen for synthetic lethal and multicopy suppressor mutants to identify two new genes involved in morphogenesis in *Saccharomyces cerevisiae*. *Mol Cell Biol* 11: 1295–305.

Beningo, K. A., Lillie, S. H., and Brown, S. S. (2000). The yeast kinesin-related protein Smy1p exerts its effects on the class V myosin Myo2p via a physical interaction. *Mol Biol Cell* 11: 691–702.

Benzer, S. (1955). Fine structure of a bacteriophage. *Proc Natl Acad Sci USA* 41: 344–54.

Benzer, S. (1962). The fine structure of the gene. *Sci Am* 206: 70–84.

Berger, J., Suzuki, T., Senti, K. A., Stubbs, J., Schaffner, G., and Dickson, B. J. (2001). Genetic mapping with SNP markers in *Drosophila*. *Nat Genet* 29: 475–81.

Betzner, A. S., Oakes, M. P., and Huttner, E. (1997). Transfer RNA-mediated suppression of amber stop codons in transgenic *Arabidopsis thaliana*. *Plant J* 11: 587–95.

Bickel, S. E., Wyman, D. W., and Orr-Weaver, T. L. (1997). Mutational analysis of the *Drosophila* sister-chromatid cohesion protein ORD and its role in the maintenance of centromeric cohesion. *Genetics* 146: 1319–31.

Bishop, D. K., Nikolski, Y., Oshiro, J., Chon, J., Shinohara, M., and Chen, X. (1999). High copy number suppression of the meiotic arrest caused by a dmc1 mutation: REC114 imposes an early recombination block and RAD54 promotes a DMC1-independent DSB repair pathway. *Genes Cells* 4: 425–44.

Blochlinger, K., Bodmer, R., Jack, J., Jan, L. Y., and Jan, Y. N. (1988). Primary structure and expression of a product from cut, a locus involved in specifying sensory organ identity in *Drosophila*. *Nature* 333: 629–35.

Borts, R. H. and Haber, J. E. (1987). Meiotic recombination in yeast: alteration by multiple heterozygosities. *Science* 237: 1459–65.

Botstein, D. and Maurer, R. (1982). Genetic approaches to the analysis of microbial development. *Ann Rev Genet* 16: 61–83.

Brachmann, R. K., Yu, K., Eby, Y., Pavletich, N. P., and Boeke, J. D. (1998). Genetic selection of intragenic suppressor mutations that reverse the effect of common p53 cancer mutations. *EMBO J* 17: 1847–59.

Bridges, C. B. (1914). Direct proof through nondisjunction that the sex-linked genes of *Drosophila* are borne on the *X* chromosome. *Science* 40: 107–9.

Bridges, C. B. (1916). Nondisjunction as proof of the chromosome theory of heredity. *Genetics* 1: 1–52, 107–63.

Bryant, P. J. and Schneiderman, H. A. (1969). Cell lineage, growth, and determination in the imaginal leg discs of *Drosophila melanogaster*. *Dev Biol* 20: 263–90.

Buckingham, R. H. (1994). Codon context and protein synthesis: enhancements of the genetic code. *Biochimie* 76: 351–4.

Buonomo, S. B., Clyne, R. K., Fuchs, J., Loidl, J., Uhlmann, F., and Nasmyth, K. (2000). Disjunction of homologous chromosomes in meiosis I depends on proteolytic cleavage of the meiotic cohesin Rec8 by separin. *Cell* 103: 387–98.

Burghes, A. H., Vaessin, H. E., and de La Chapelle, A. (2001). Genetics. The land between Mendelian and multifactorial inheritance. *Science* 293: 2213–14.

Buvoli, M., Buvoli, A., and Leinwand, L. A. (2000). Suppression of nonsense mutations in cell culture and mice by multimerized suppressor tRNA genes. *Mol Cell Biol* 20: 3116–24.

Cairns, B. R., Henry, N. L., and Kornberg, R. D. (1996). TFG/TAF30/ANC1, a component of the yeast SWI/SNF complex that is similar to the leukemogenic proteins ENL and AF-9. *Mol Cell Biol* 16: 3308–16.

Cao, L., Alani, E., and Kleckner, N. (1990). A pathway for generation and processing of double-strand breaks during meiotic recombination in *S. cerevisiae*. *Cell* 61: 1089–101.

Caplen, N. J., Fleenor, J., Fire, A., and Morgan, R. A. (2000). dsRNA-mediated gene silencing in cultured Drosophila cells: a tissue culture model for the analysis of RNA interference. *Gene* 252: 95–105.

Carlson, M. (1997). Genetics of transcriptional regulation in yeast: connections to the RNA polymerase II CTD. *Ann Rev Cell Dev Biol* 13: 1–23.

Carlson, P. (1971). A genetic analysis of the *rudimentary* locus of *Drosophila melanogaster*. *Genet Res (Camb)* 17: 53–81.

Carpenter, A. T. (1973). A meiotic mutant defective in distributive disjunction in *Drosophila melanogaster*. *Genetics* 73: 393–428.

Carpenter, A. T. (1975). Electron microscopy of meiosis in *Drosophila melanogaster* females: II. The recombination nodule – a recombination-associated structure at pachytene? *Proc Natl Acad Sci USA* 72: 3186–9.

Carpenter, A. T. (1979a). Synaptonemal complex and recombination nodules in wild-type *Drosophila melanogaster* females. *Genetics* 92: 511–41.

Carpenter, A. T. (1979b). Recombination nodules and synaptonemal complex in recombination-defective females of *Drosophila melanogaster*. *Chromosoma* 75: 259–92.

Carpenter, A. T. (1981). EM autoradiographic evidence that DNA synthesis occurs at recombination nodules during meiosis in *Drosophila melanogaster* females. *Chromosoma* 83: 59–80.

Carpenter, A. T. (1984). Recombination nodules and the mechanism of crossing-over in *Drosophila*. *Symp Soc Exp Biol* 38: 233–43.

Carpenter, A. T. (1987). Gene conversion, recombination nodules, and the initiation of meiotic synapsis. *Bioessays* 6: 232–6.

Carpenter, A. T. (1988). Thoughts on recombination nodules, meiotic recombination, and chiasmata. In: R. Kucherlapati (ed.), *Genetic Recombination*. ASM Press, Washington, DC, pp. 529–49.

Carthew, R. W. (2001). Gene silencing by double-stranded RNA. *Curr Opin Cell Biol* 13: 244–8.

Carthew, R. W. and Rubin, G. M. (1990). *seven in absentia*, a gene required for specification of R7 cell fate in the *Drosophila* eye. *Cell* 63: 561–77.

Case, M. E. and Giles, N. H. (1958). Evidence from tetrad analysis for both normal and abnormal recombination between allelic mutants in *Neurospora crassa*. *Proc Natl Acad Sci USA* 44: 378–90.

Case, M. E. and Giles, N. H. (1959). Recombination mechanisms at the pan-3 locus in *Neurospora crassa*. *Cold Spring Harbor Symp Quant Biol* 23: 119–35.

Case, M. E. and Giles, N. H. (1964). Allelic recombination in *Neurospora*: tetrad analysis of a three point cross within the pan-2 locus. *Genetics* 49: 529–40.

Chen, Z., Ulmasov, B., and Folk, W. R. (1998). Nonsense and missense translational suppression in plant cells mediated by tRNA(Lys). *Plant Mol Biol* 36: 163–70.

Chikashige, Y., Ding, D. Q., Funabiki, H., Haraguchi, T., Mashiko, S., Yanagida, M., and Hiraoka, Y. (1994). Telomere-led premeiotic chromosome movement in fission yeast. *Science* 264: 270–73.

Chikashige, Y., Ding, D. Q., Imai, Y., Yamamoto, M., Haraguchi, T., and Hiraoka, Y. (1997). Meiotic nuclear reorganization: switching the position of centromeres and telomeres in the fission yeast *Schizosaccharomyces pombe*. *EMBO J* 16: 193–202.

Chovnick, A. (1989). Intragenic recombination in *Drosophila*: the *rosy* locus. *Genetics* 123: 621–4.

Chua, P. R. and Roeder, G. S. (1997). Tam1, a telomere-associated meiotic protein, functions in chromosome synapsis and crossover interference. *Genes Dev* 11: 1786–800.

Chuang, C. F. and Meyerowitz, E. M. (2000). Specific and heritable genetic interference by double-stranded RNA in Arabidopsis thaliana. *Proc Natl Acad Sci USA* 97: 4985–90.

Clarke, L. (1998). Centromeres: proteins, protein complexes, and repeated domains at centromeres of simple eukaryotes. *Curr Opin Genet Dev* 8: 212–18.

Clarke, L. and Carbon, J. (1980). Isolation of a yeast centromere and construction of functional small circular chromosomes. *Nature* 287: 504–9.

Clarke, L. and Carbon, J. (1985). The structure and function of yeast centromeres. *Ann Rev Genet* 19: 29–55.

Clarke, L., Baum, M., Marschall, L. G., Ngan, V. K., and Steiner, N. C. (1993). Structure and function of *Schizosaccharomyces pombe* centromeres. *Cold Spring Harbor Symp Quant Biol* 58: 687–95.

Cline, T. W. (1978). Two closely linked mutations in *Drosophila melanogaster* that are lethal to opposite sexes and interact with *daughterless*. *Genetics* 90: 683–98.

Cline, T. W. (1984). Autoregulation functioning of a *Drosophila* gene product that establishes and maintains the sexually determined state. *Genetics* 107: 231–77.

Cline, T. W. and Meyer, B. J. (1996). Vive la différence: males vs females in flies vs worms. *Ann Rev Genet* 30: 637–702.

Conrad, M. N., Dominguez, A. M., and Dresser, M. E. (1997). Ndj1p, a meiotic telomere protein required for normal chromosome synapsis and segregation in yeast. *Science* 276: 1252–5.

Cooper, J. P., Watanabe, Y., and Nurse, P. (1998). Fission yeast Taz1 protein is required for meiotic telomere clustering and recombination. *Nature* 392: 828–31.

Cooper, K. W. (1964). Meiotic conjunctive elements not involving chiasmata. *Proc Natl Acad Sci USA* 52: 1248–55.

Copeland, N. G., Jenkins, N. A., and Court, D. L. (2001). Recombineering: a powerful new tool for mouse functional genomics. *Nat Rev Genet* 2: 769–79.

Copenhaver, G. P., Browne, W. E., and Preuss, D. (1998). Assaying genome-wide recombination and centromere functions with *Arabidopsis* tetrads. *Proc Natl Acad Sci USA* 95: 247–52.

Copenhaver, G. P., Keith, K. C., and Preuss, D. (2000). Tetrad analysis in higher plants. A budding technology. *Plant Physiol* 124: 7–16.

Craymer, L. (1981). Techniques for manipulating chromosomal rearrangements and their application to *Drosophila melanogaster*. I. Pericentric inversions. *Genetics* 99: 75–97.

Craymer, L. (1984). Techniques for manipulating chromosomal rearrangements and their application to *Drosophila melanogaster*. II. Translocations. *Genetics* 108: 573–87.

Creighton, H. B. and McClintock, B. (1931). A correlation of cytological and genetical crossing-over in *Zea mays*. *Proc Natl Acad Sci USA* 17: 492–7.

Crick, F. H., Barnett, L., Brenner, S., and Watts-Tobin, R. J. (1961). General nature of the genetic code for proteins. *Nature* 192: 1227–332.

Cromie, G. A. and Leach, D. R. (2000). Control of crossing over. *Mol Cell* 6: 815–26.

Cullen, C. F., May, K. M., Hagan, I. M., Glover, D. M., and Ohkura, H. (2000). A new genetic method for isolating functionally interacting genes: high plo1(+)-dependent mutants and their suppressors define genes in mitotic and septation pathways in fission yeast. *Genetics* 155: 1521–34.

Davis, L. and Smith, G. R. (2001). Meiotic recombination and chromosome segregation in *Schizosaccharomyces pombe*. *Proc Natl Acad Sci USA* 98: 8395–402.

Dawson, D. S., Murray, A. W., and Szostak, J. W. (1986). An alternative pathway for meiotic chromosome segregation in yeast. *Science* 234: 713–17.

Denell, R. E. (1972). The nature of reversion of a dominant gene of *Drosophila melanogaster*. *Mutat Res* 15: 221–3.

Dernburg, A. F., Sedat, J. W., Cande, W. Z., and Bas, H. W. (1995). Cytology of telomeres. In: E. H. Blackburn and C. W. Greider (eds.), *Telomeres*. Cold Spring Harbor Press, New York, pp. 295–338.

Dernburg, A. F., Sedat, J. W., and Hawley, R. S. (1996). Direct evidence of a role for heterochromatin in meiotic chromosome segregation. *Cell* 86: 135–46.

Dernburg, A. F., McDonald, K., Moulder, G., Barstead, R., Dresser, M., and Villeneuve, A. M. (1998). Meiotic recombination in *C. elegans* initiates by a conserved mechanism and is dispensable for homologous chromosome synapsis. *Cell* 94: 387–98.

Dernburg, A. F., Zalevsky, J., Colaiacovo, M. P., and Villeneuve, A. M. (2000). Transgene-mediated cosuppression in the *C. elegans* germ line. *Genes Dev* 14: 1578–83.

di Rago, J. P., Hermann-Le Denmat, S., Paques, F., Risler, F. P., Netter, P., and Slonimski, P. P. (1995). Genetic analysis of the folded structure of yeast mitochondrial cytochrome b by selection of intragenic second-site revertants. *J Mol Biol* 248: 804–11.

Doheny, K. F., Sorger, P. K., Hyman, A. A., Tugendreich, S., Spencer, F., and Hieter, P. (1993). Identification of essential components of the S. cerevisiae kinetochore. *Cell* 73: 761–74.

Dorsett, D. (1993). Distance-independent inactivation of an enhancer by the suppressor of Hairy-wing DNA-binding protein of *Drosophila*. *Genetics* 134: 1135–44.

Drubin, D. G., Miller, K. G., and Botstein, D. (1988). Yeast actin-binding proteins: evidence for a role in morphogenesis. *J Cell Biol* 107: 2551–61.

Duncan, I. W. and Kaufman, T. C. (1975). Cytogenic analysis of chromosome 3 in *Drosophila melanogaster*: mapping of the proximal portion of the right arm. *Genetics* 80: 733–52.

Dutcher, S. K. (1995). Mating and tetrad analysis in *Chlamydomonas renhardtii*. *Methods Cell Biol* 47: 531–40.

Edgley, M. L., Baillie, D. L., Riddle, D. L., and Rose, A. M. (1995). Genetic balancers. *Methods Cell Biol* 48: 147–84.

Elkins, T., Zinn, K., McAllister, L., Hoffmann, F. M., and Goodman, C. S. (1990). Genetic analysis of a *Drosophila* neural cell adhesion molecule: interaction of fasciclin I and Abelson tyrosine kinase mutations. *Cell* 60: 565–75.

Ephrussi, B. and Beadle, G. W. (1936). A technique for transportation for *Drosophila*. *Am Nat* 70: 218–25.

Erdman, S. E., Chen, H. J., and Burtis, K. C. (1996). Functional and genetic characterization of the oligomerization and DNA binding properties of the *Drosophila* doublesex proteins. *Genetics* 144: 1639–52.

Escobar, M. A., Civerolo, E. L., Summerfelt, K. R., and Dandekar, A. M. (2001). RNAi-mediated oncogene silencing confers resistance to crown gall tumorigenesis. *Proc Natl Acad Sci USA* 98: 13437–42.

Farrall, M. (1997). Affected sibpair linkage tests for multiple linked susceptibility genes. *Genet Epidemiol* 14: 103–15.

Ferguson, E. L., Sternberg, P. W., and Horvitz, H. R. (1987). A genetic pathway for the specification of the vulval cell lineages of *Caenorhabditis elegans*. *Nature* 326: 259–67.

Fincham, J. (1966). *Genetic Complementation*. Benjamin, New York.

Fink, G. R. (1966). A cluster of genes controlling three enzymes in histidine biosynthesis in *Saccharomyces cerevisiae*. *Genetics* 53: 445–9.

Finnerty, V. (1976). Genetic units in Drosophila: simple cistrons segregation. In: M. Ashburner and E. Novitski (eds.), *The Genetics and Biology of Drosophila*. Vol. 1a. Academic Press, London, pp. 721–65.

Fogel, S. and Mortimer, R. K. (1969). Informational transfer in meiotic gene conversion. *Proc Natl Acad Sci USA* 62: 96–103.

Fogel, S. and Mortimer, R. K. (1970). Fidelity of gene conversion in yeast. *Mol Gen Genet* 109: 177–86.

Fogel, S., Mortimer, R. K., and Lusnak, K. (1982). Mechanisms of meiotic gene conversion, or "wanderings on a foreign strand." In: J. H. Strathern and E. W. Jones Broach Jr. (eds.), *The Molecular Biology of the Yeast Saccharomyces 1*. Cold Spring Harbor Press, New York, pp. 289–339.

Forsburg, S. L. (2001). The art and design of genetic screens: yeast. *Nat Rev Genet* 2: 659–68.

Foss, E. J. and Stahl, F. W. (1995). A test of a counting model for chiasma interference. *Genetics* 139: 1201–9.

Foss, E., Lande, R., Stahl, F. W., and Steinberg, C. M. (1993). Chiasma interference as a function of genetic distance. *Genetics* 133: 681–91.

Freund, J. N. and Jarry, B. P. (1987). The rudimentary gene of Drosophila melanogaster encodes four enzymic functions. *J Mol Biol* 193: 1–13.

Freund, J. N., Zerges, W., Schedl, P., Jarry, B. P., and Vergis, W. (1986). Molecular organization of the *rudimentary* gene of *Drosophila melanogaster*. *J Mol Biol* 189: 25–36.

Fridell, R. A. and Searles, L. L. (1994). Evidence for a role of the *Drosophila melanogaster* suppressor of sable gene in the pre-mRNA splicing pathway. *Mol Cell Biol* 14: 859–67.

Fridell, R. A., Pret, A. M., and Searles, L. L. (1990). A retrotransposon 412 insertion within an exon of the *Drosophila melanogaster vermilion* gene is spliced from the precursor RNA. *Genes Dev* 4: 559–66.

Frischer, L. E., Hagen, F. S., and Garber, R. L. (1986). An inversion that disrupts the *Antennapedia* gene causes abnormal structure and localization of RNAs. *Cell* 47: 1017–23.

Fuller, M. T. (1986). Genetic analysis of spermatogenesis in *Drosophila*: the role of the testis-specific beta-tubulin and interacting genes in cellular morphogenesis. In: J. G. Gall (ed.), *Gametogenesis and the Early Embryo*. Alan R. Liss, New York, pp. 19–42.

Game, J. C., Sitney, K. C., Cook, V. E., and Mortimer, R. K. (1989). Use of a ring chromosome and pulsed-field gels to study interhomolog recombination, double-strand DNA breaks and sister-chromatid exchange in yeast. *Genetics* 123: 695–713.

Ganetzky, B. (1977). On the components of segregation distortion in *Drosophila melanogaster*. *Genetics* 86: 321–55.

Garcia-Bellido, A. and Merriam, J. R. (1969). Cell lineage of the imaginal discs in *Drosophila* gynandromorphs. *J Exp Zool* 170: 61–76.

Garza, D., Medhora, M. M., and Hartl, D. L. (1990). Drosophila nonsense suppressors: functional analysis in *Saccharomyces cerevisiae*, *Drosophila* tissue culture cells and *Drosophila melanogaster*. *Genetics* 126: 625–37.

Gelbart, W. M. (1974). A new mutant controlling mitotic chromosome disjunction in *Drosophila melanogaster*. *Genetics* 76: 51–63.

Gelbart, W. M. (1982). Synapsis-dependent allelic complementation at the decapentaplegic gene complex in *Drosophila melanogaster*. *Proc Natl Acad Sci USA* 79: 2636–40.

Gepner, J., Li, M., Ludmann, S., Kortas, C., Boylan, K., Iyadurai, S. J., McGrail, M., and Hays, T. S. (1996). Cytoplasmic dynein function is essential in *Drosophila melanogaster*. *Genetics* 142: 865–78.

Gesteland, R. F., Wolfner, M., Grisafi, P., Fink, G., Botstein, D., and Roth, J. R. (1976). Yeast suppressors of UAA and UAG nonsense codons work efficiently in vitro via tRNA. *Cell* 7: 381–90.

Geyer, P. K. and Corces, V. G. (1987). Separate regulatory elements are responsible for the complex pattern of tissue-specific and developmental transcription of the *yellow* locus in *Drosophila melanogaster*. *Genes Dev* 1: 996–1004.

Geyer, P. K., Spana, C., and Corces, V. G. (1986). On the molecular mechanism of gypsy-induced mutations at the *yellow* locus of *Drosophila melanogaster*. *EMBO J* 5: 2657–62.

Geyer, P. K., Green, M. M., and Corces, V. G. (1990). Tissue-specific transcriptional enhancers may act in trans on the gene located in the homologous chromosome: the molecular basis of transvection in *Drosophila*. *EMBO J.* 9: 2247–56.

Gibson, G. and Hogness, D. S. (1996). Effect of polymorphism in the *Drosophila* regulatory gene *Ultrabithorax* on homeotic stability. *Science* 271: 200–203.

Goldberg, A. F. and Molday, R. S. (1996). Defective subunit assembly underlies a digenic form of retinitis pigmentosa linked to mutations in peripherin/rds and rom-1. *Proc Natl Acad Sci USA* 93: 13726–30.

Golic, K. G. (1991). Site-specific recombination between homologous chromosomes in *Drosophila*. *Science* 252: 958–61.

Golic, K. G. (1993). Generating mosaics by site-specific recombination. In: D. A. Hartley (ed.), *Cellular Interactions in Development: A Practical Approach*. IRL Press, Oxford, pp. 1–32.

Golic, M. M. and Golic, K. G. (1996). A quantitative measure of the mitotic pairing of alleles in *Drosophila melanogaster* and the influence of structural heterozygosity. *Genetics* 143: 385–400.

Green, L. L., Wolf, N., McDonald, K. L., and Fuller, M. T. (1990). Two types of genetic interaction implicate the *whirligig* gene of *Drosophila melanogaster* in microtubule organization in the flagellar axoneme. *Genetics* 126: 961–73.

Green, M. M. (1963). Interallelic complementation and recombination at the *rudimentary* wing locus in *Drosophila*. *Genetica* 34: 242–53.

Green, M. M. (1977). X-ray induced reversions of the mutant forked-3N in *Drosophila melanogaster*, a reappraisal. *Mutat Res* 43: 305–8.

Green, M. M. (1990). The foundations of genetic fine structure: a retrospective from memory. *Genetics* 124: 793–6.

Green, M. M. and Green, K. C. (1949). Crossing over between alleles at the *lozenge* locus in *Drosophila melanogaster*. *Proc Natl Acad Sci USA* 35: 586–91.

Greenspan, R. J. (1997). *Fly Pushing: The Theory and Practice of Drosophila Genetics*. Cold Spring Harbor Laboratory Press, New York.

Grell, E. H. (1964). Influence of the location of a chromosomal duplication on crossingover in *Drosophila melanogaster* (abstract). *Genetics* 50: 251–2.

Grell, R. F. (1976). Distributive pairing segregation. In: M. Ashburner and E. Novitski (eds.), *The Genetics and Biology of Drosophila*. Vol. 1a. Academic Press, London, pp. 436–86.

Grelon, M., Vezon, D., Gendrot, G., and Pelletier, G. (2001). AtSPO11-1 is necessary for efficient meiotic recombination in plants. *EMBO J* 20: 589–600.

Grishok, A., Pasquinelli, A. E., Conte, D., Li, N., Parrish, S., Ha, I., Baillie, D. L., Fire, A., Ruvkun, G., and Mello, C. C. (2001). Genes and mechanisms related to RNA interference regulate expression of the small temporal RNAs that control *C. elegans* developmental timing. *Cell* 106: 23–34.

Guacci, V. and Kaback, D. B. (1991). Distributive disjunction of authentic chromosomes in *Saccharomyces cerevisiae*. *Genetics* 127: 475–88.

Guarente, L. (1993). Synthetic enhancement in gene interaction: a genetic tool come of age. *Trends Genet* 9: 362–6.

Guthrie, C. and Fink, G. R. (1991). *Guide to Yeast Genetics and Molecular Biology*. Academic Press, New York.

Haber, J. E., Thorburn, P. C., and Rogers, D. (1984). Meiotic and mitotic behavior of dicentric chromosomes in *Saccharomyces cerevisiae*. *Genetics* 106: 185–205.

Haldane, J. B. S. (1919). *J Genet* 8: 299–309.

Hall, J. C. (1979). Control of male reproductive behavior by the central nervous system of Drosophila: dissection of a courtship pathway by genetic mosaics. *Genetics* 92: 437–57.

Hall, J. C., Gelbart, W. M., and Kankel, D. R. (1976). Mosaic systems. Segregation. In: M. Ashburner and E. Novitski (eds.), *The Genetics and Biology of Drosophila*. Vol. 1a. Academic Press, London, pp. 265–314.

Halsell, S. R. and Kiehart, D. P. (1998). Second-site noncomplementation identifies genomic regions required for Drosophila nonmuscle myosin function during morphogenesis. *Genetics* 148: 1845–63.

Halsell, S. R., Chu, B. I., and Kiehart, D. P. (2000). Genetic analysis demonstrates a direct link between rho signaling and nonmuscle myosin function during *Drosophila* morphogenesis. *Genetics* 155: 1253–65.

Hanein, D., Volkmann, N., Goldsmith, S., Michon, A. M., Lehman, W., Craig, R., DeRosier, D., Almo, S., and Matsudaira, P. (1998). An atomic model of fimbrin binding to F-actin and its implications for filament crosslinking and regulation. *Nat Struct Biol* 5: 787–92.

Harrison, D. A., Geyer, P. K., Spana, C., and Corces, V. G. (1989). The *gypsy* retrotransposon of *Drosophila melanogaster*: mechanisms of mutagenesis and interaction with the suppressor of *Hairy-wing* locus. *Dev Genet* 10: 239–48.

Hartman, J. L. t., Garvik, B., and Hartwell, L. (2001). Principles for the buffering of genetic variation. *Science* 291: 1001–4.

Hartman, P. E. and Roth, J. R. (1973). Mechanisms of suppression. *Adv Genet* 17: 1–105.

Hartwell, L. H. (1991). Twenty-five years of cell cycle genetics. *Genetics* 129: 975–80.

Hartwell, L. H., Culotti, J., and Reid, B. (1970). Genetic control of the cell-division cycle in yeast. I. Detection of mutants. *Proc Natl Acad Sci USA* 66: 352–9.

Hartwell, L. H., Mortimer, R. K., Culotti, J., and Culotti, M. (1973). Genetic control of the cell division cycle in yeast. V. Genetic analysis of *cdc* mutants. *Genetics* 74: 267–86.

Hassold, T. and Hunt, P. (2001). To err (meiotically) is human: the genesis of human aneuploidy. *Nat Rev Genet* 2: 280–91.

Hassold, T., Merrill, M., Adkins, K., Freeman, S., and Sherman, S. (1995). Recombination and maternal age-dependent nondisjunction: molecular studies of trisomy 16. *Am J Hum Genet* 57: 867–74.

Hawley, R. S. (1980). Chromosomal sites necessary for normal levels of meiotic recombination in *Drosophila melanogaster*. I. Evidence for and mapping of the sites. *Genetics* 94: 625–46.

Hawley, R. S. (1988). Exchange and chromosome segregation in eukaryotes. In: R. Kucherlapati (ed.), *Genetic Recombination*. ASM Press, Washington, DC, pp. 497–528.

Hawley, R. S. (1993). Meiosis as an "M" thing: twenty-five years of meiotic mutants in *Drosophila*. *Genetics* 135: 613–18.

Hawley, R. S. and Theurkauf, W. E. (1993). Requiem for the distributive system: achiasmate segregation in *Drosophila* females. *Trends Genet* 9: 310–17.

Hawley, R. S., Irick, H., Zitron, A. E., et al. (1992). There are two mechanisms of achiasmate segregation in *Drosophila* females, one of which requires heterochromatic homology. *Dev Genet* 13: 440–67.

Hawley, R. S., McKim, K. S., and Arbel, T. (1993). Meiotic segregation in *Drosophila melanogaster* females: molecules, mechanisms, and myths. *Ann Rev Genet* 27: 281–317.

Hawthorne, D. C. and Leupold, U. (1974). Suppressors in yeast. *Curr Top Microbiol Immunol* 64: 1–47.

Hays, T. S., Deuring, R., Robertson, B., Prout, M., and Fuller, M. T. (1989). Interacting proteins identified by genetic interactions: a missense mutation in alpha-tubulin fails to complement alleles of the testis-specific beta-tubulin gene of *Drosophila melanogaster*. *Mol Cell Biol* 9: 875–84.

Hazelrigg, T. and Kaufman, T. C. (1983). Revertants of dominant mutations associated with the Antennapedia gene complex of *Drosophila melanogaster*: cytology and genetics. *Genetics* 105: 581–600.

Hendrix, R. W. (1983). *Lambda II*. Cold Spring Harbor Laboratory Press, Cold Spring Harbor, NY.

Hereford, L. M. and Hartwell, J. H. (1974). Sequential gene function in the initiation of *Saccharomyces cerevisiae* DNA synthesis. *J Mol Biol* 85: 445–61.

Herman, R. K. (1984). Analysis of genetic mosaics of the nematode *Caneorhabditis elegans*. *Genetics* 108: 165–80.

Herman, R. K. and Kari, C. K. (1989). Recombination between small X chromosome duplications and the X chromosome in *Caenorhabditis elegans*. *Genetics* 121: 723–37.

Hinton, C. W. (1955). The behavior of an unstable ring chromosome of *Drosophila melanogaster*. *Genetics* 40: 951–61.

Hinton, C. W. (1957). The analysis of rod derivatives of an unstable ring chromosome in *Drosophila melanogaster*. *Genetics* 42: 55–65.

Hipeau-Jacquotte, R., Brutlag, D. L., and Bregegere, F. (1989). Conversion and reciprocal exchange between tandem repeats in *Drosophila melanogaster*. *Mol Gen Genet* 220: 140–6.

Hoff, T., Schnorr, K. M., and Mundy, J. (2001). A recombinase-mediated transcriptional induction system in transgenic plants. *Plant Mol Biol* 45: 41–9.

Hoja, U., Wellein, C., Greiner, E., and Schweizer, E. (1998). Pleiotropic phenotype of acetyl-CoA-carboxylase-defective yeast cells – viability of a BPL1-amber mutation depending on its readthrough by normal tRNA(Gln) (CAG). *Eur J Biochem* 254: 520–26.

Holliday, R. (1964). A mechanism for gene conversion in fungi. *Genet Res (Camb)* 5: 282–304.

Holm, D. G. (1976). Compound autosomes. In: M. Ashburner and E. Novitski (eds.), *The Genetics and Biology of Drosophila*. Vol. 1b. Academic Press, London, pp. 529–61.

Hong, S. and Spreitzer, R. J. (1997). Complementing substitutions at the bottom of the barrel influence catalysis and stability of ribulose-bisphosphate carboxylase/oxygenase. *J Biol Chem* 272: 11114–17.

Honts, J. E., Sandrock, T. S., Brower, S. M., O'Dell, J. L., and Adams, A. E. (1994). Actin mutations that show suppression with fimbrin mutations identify a likely fimbrin-binding site on actin. *J Cell Biol* 126: 413–22.

Hope, I. A. (2001). RNAi surges on: application to cultured mammalian cells. *Trends Genet* 17: 440.

Hoppe, P. E. and Greenspan, R. J. (1986). Local function of the Notch gene for embryonic ectodermal pathway choice in Drosophila. *Cell* 46: 773–83.

Hsieh, J. and Fire, A. (2000). Recognition and silencing of repeated DNA. *Ann Rev Genet* 34: 187–204.

Hunt, P. A. and LeMaire-Adkins, R. (1998). Genetic control of mammalian female meiosis. *Curr Top Dev Biol* 37: 359–81.

Hunter, C. P. (2000). Gene silencing: shrinking the black box of RNAi. *Curr Biol* 10: R137–40.

Hurst, D. D., Fogel, S., and Mortimer, R. K. (1972). Conversion-associated recombination in yeast (hybrids-meiosis-tetrads-marker loci-models). *Proc Natl Acad Sci USA* 69: 101–5.

Hyman, A. A. and Sorger, P. K. (1995). Structure and function of kinetochores in budding yeast. *Ann Rev Cell Dev Biol* 11: 471–95.

Jack, J. W. (1985). Molecular organization of the *cut* locus of *Drosophila melanogaster*. *Cell* 42: 869–76.

Jack, J. and DeLotto, Y. (1995). Structure and regulation of a complex locus: the cut gene of *Drosophila*. *Genetics* 139: 1689–700.

Jack, J. W. and Judd, B. H. (1979). Allelic pairing and gene regulation: a model for the *zeste–white* interaction. *Proc Natl Acad Sci* 76: 1368–72.

Jackson, S. M. and Berg, C. A. (1999). Soma-to-germline interactions during *Drosophila* oogenesis are influenced by dose-sensitive interactions between *cut* and the genes *cappuccino*, *ovarian tumor* and *agnostic*. *Genetics* 153: 289–303.

Janssens, F. A. (1909). Spermatogénèse dans les Batraciens. V. La théorie de las chiasmatypie. Nouvelles interprétation des cinèses de maturation. *Cellule* 25: 387–411.

Jarvik, J. and Botstein, D. (1975). Conditional-lethal mutations that suppress genetic defects in morphogenesis by altering structural proteins. *Proc Natl Acad Sci USA* 72: 2738–42.

Jiang, Y., Scarpa, A., Zhang, L., Stone, S., Feliciano, E., and Ferro-Novick, S. (1998). A high copy suppressor screen reveals genetic interactions between BET3 and a new gene. Evidence for a novel complex in ER-to-Golgi transport. *Genetics* 149: 833–41.

Jinks-Robertson, S. and Petes, T. D. (1985). High-frequency meiotic gene conversion between repeated genes on non-homologous chromosomes in yeast. *Proc Natl Acad Sci USA* 82: 3350–4.

Jinks-Robertson, S. and Petes, T. D. (1986). Chromosomal translocations generated by high-frequency meiotic recombination between repeated yeast genes. *Genetics* 114: 731–52.

Johnson, S. L., Africa, D., Horne, S., and Postlethwait, J. H. (1995). Half-tetrad analysis in zebrafish: mapping the ros mutation and the centromere of linkage group I. *Genetics* 139: 1727–35.

Johnson, T. K. and Judd, B. H. (1979). Analysis of the *cut* locus in *Drosophila melanogaster*. *Genetics* 92: 485–502.

Jones, G. H. (1984). The control of chiasma distribution. In: C. W. Evans and H. G. Dickinson (eds.), *Controlling*

Events in Meiosis. The Company of Biologists, Cambridge, pp. 293–320.

Judd, B. H. (1976). Genetics units of Drosophila – complex loci. In: M. Ashburner and E. Novitski (eds.), *The Genetics and Biology of Drosophila*. Vol. 1a. Academic Press, London, pp. 767–99.

Jung, K. H. and Spudich, J. L. (1998). Suppressor mutation analysis of the sensory rhodopsin I-transducer complex: insights into the color-sensing mechanism. *J Bacteriol* 180: 2033–42.

Justice, M. J. (2000). Capitalizing on large-scale mouse mutagenesis screens. *Nat Rev Genet* 1: 109–15.

Kaback, D. B. (1989). Meiotic segregation of circular plasmid-minichromosomes from intact chromosomes in *Saccharomyces cerevisiae*. *Curr Genet* 15: 385–92.

Kajiwara, K., Berson, E. L., and Dryja, T. P. (1994). Digenic retinitis pigmentosa due to mutations at the unlinked peripherin/RDS and ROM1 loci. *Science* 264: 1604–8.

Kang, M. E. and Dahmus, M. E. (1995). The unique C-terminal domain of RNA polymerase II and its role in transcription. *Adv Enzymol Relat Areas Mol Biol* 71: 41–77.

Karim, F. D., Chang, H. C., Therrien, M., Wassarman, D. A., Laverty, T., and Rubin, G. M. (1996). A screen for genes that function downstream of Ras1 during *Drosophila* eye development. *Genetics* 143: 315–29.

Karpen, G. H. and Allshire, R. C. (1997). The case for epigenetic effects on centromere identity and function. *Trends Genet* 13: 489–96.

Karpen, G. H., Le, M. H., and Le, H. (1996). Centric heterochromatin and the efficiency of achiasmate disjunction in *Drosophila* female meiosis. *Science* 273: 118–22.

Katsanis, N., Lupski, J. R., and Beales, P. L. (2001). Exploring the molecular basis of Bardet–Biedl syndrome. *Hum Mol Genet* 10: 2293–9.

Keesey Jr., J. K., Bigelis, R., and Fink, G. R. (1979). The product of the *his4* gene cluster in *Saccharomyces cerevisiae*. A trifunctional polypeptide. *J Biol Chem* 254: 7427–33.

Kemphues, K. J. R., Elizabeth, C., Raff, R. A., and Kaufman, T. C. (1980). Mutation in a testis-specific beta-tubulin in *Drosophila*: analysis of its effects on meiosis and map location of the gene. *Cell* 21: 445–51.

Kemphues, K. J. R., Elizabeth, C., and Kaufman, T. C. (1983). Genetic analysis of *B2t*, the structural gene for a testis-specific beta-tubulin subunit in *Drosophila melanogaster*. *Genetics* 105: 345–56.

Kennerdell, J. R. and Carthew, R. W. (1998). Use of dsRNA-mediated genetic interference to demonstrate that frizzled and frizzled 2 act in the wingless pathway. *Cell* 95: 1017–26.

Kennerdell, J. R. and Carthew, R. W. (2000). Heritable gene silencing in Drosophila using double-stranded RNA. *Nat Biotechnol* 18: 896–8.

Kidd, T., Bland, K. S., and Goodman, C. S. (1999). Slit is the midline repellent for the robo receptor in *Drosophila*. *Cell* 96: 785–94.

Kim, D. W., Sacher, M., Scarpa, A., Quinn, A. M., and Ferro-Novick, S. (1999). High-copy suppressor analysis reveals a physical interaction between Sec34p and Sec35p, a protein implicated in vesicle docking. *Mol Biol Cell* 10: 3317–29.

Kim, N., Kim, J., Park, D., Rosen, C., Dorsett, D., and Yim, J. (1996). Structure and expression of wild-type and suppressible alleles of the *Drosophila purple* gene. *Genetics* 142: 1157–68.

Kimble, J. and Austin, J. (1989). Genetic control of cellular interactions in *Caenorhabditis elegans* development. *Ciba Found Symp* 144: 212–20; discussion 221–6, 290–5.

Kirkpatrick, D. T., Wang, Y. H., Dominska, M., Griffith, J. D., and Petes, T. D. (1999). Control of meiotic recombination and gene expression in yeast by a simple repetitive DNA sequence that excludes nucleosomes. *Mol Cell Biol* 19: 7661–71.

Klar, A. J., Strathern, J. N., and Abraham, J. A. (1984). Involvement of double-strand chromosomal breaks for mating-type switching in *Saccharomyces cerevisiae*. *Cold Spring Harbor Symp Quant Biol* 49: 77–88.

Knowles, B. A. and Hawley, R. S. (1991). Genetic analysis of microtubule motor proteins in *Drosophila*: a mutation at the *ncd* locus is a dominant enhancer of *nod*. *Proc Natl Acad Sci USA* 88: 7165–9.

Koehler, K. E., Boulton, C. L., Collins, H. E., French, R. L., Herman, K. C., Lacefield, S. M., Madden, L. D., Schuetz, C. D., and Hawley, R. S. (1996). Spontaneous X chromosome MI and MII nondisjunction events in *Drosophila melanogaster* oocytes have different recombinational histories. *Nat Genet* 14: 406–14.

Kopczynski, J. B., Raff, A. C., and Bonner, J. J. (1992). Translational readthrough at nonsense mutations in the HSF1 gene of *Saccharomyces cerevisiae*. *Mol Gen Genet* 234: 369–78.

Kusano, A., Staber, C., and Ganetzky, B. (2001a). Nuclear mislocalization of enzymatically active RanGAP causes segregation distortion in *Drosophila*. *Dev Cell* 1: 351–61.

Kusano, K., Johnson-Schlitz, D. M., and Engels, W. R. (2001b). Sterility of *Drosophila* with mutations in the Bloom syndrome gene – complementation by Ku70. *Science* 291: 2600–2.

Lamb, N. E., Freeman, S. B., Savage-Austin, A., Pettay, D., Taft, L., Hersey, J., Gu, Y., Shen, J., Saker, D., May, K. M., Avramopoulos, D., Petersen, M. B., Hallberg, A.,

Mikkelsen, M., Hassold, T. J., and Sherman, S. L. (1996). Susceptible chiasmate configurations of chromosome 21 predispose to non-disjunction in both maternal meiosis I and meiosis II. *Nat Genet* 14: 400–5.

Lander, E. S. and Botstein, D. (1989). Mapping mendelian factors underlying quantitative traits using RFLP linkage maps. *Genetics* 121: 185–99.

Lawrie, N. M., Tease, C., and Hulten, M. A. (1995). Chiasma frequency, distribution and interference maps of mouse autosomes. *Chromosoma* 104: 308–14.

Le, M. H., Duricka, D., and Karpen, G. H. (1995). Islands of complex DNA are widespread in *Drosophila* centric heterochromatin. *Genetics* 141: 283–303.

Lee, J. Y. and Orr-Weaver, T. L. (2001). The molecular basis of sister-chromatid cohesion. *Ann Rev Cell Dev Biol* 17: 753–77.

Lefevre, G. and Watkins, W. (1986). The question of the total gene number in *Drosophila melanogaster*. *Genetics* 113: 869–95.

Lewis, E. B. (1954). The theory and application of a new method of detecting chromosomal rearrangements in *Drosophila melanogaster*. *Am Nat* 88: 225–39.

Lichten, M. (2001). Meiotic recombination: breaking the genome to save it. *Curr Biol* 11: R253–6.

Lichten, M., Borts, R. H., and Haber, J. E. (1987). Meiotic gene conversion and crossing over between dispersed homologous sequences occurs frequently in *Saccharomyces cerevisiae*. *Genetics* 115: 233–46.

Lillie, S. H. and Brown, S. S. (1992). Suppression of a myosin defect by a kinesin-related gene. *Nature* 356: 358–61.

Lillie, S. H. and Brown, S. S. (1998). Smy1p, a kinesin-related protein that does not require microtubules. *J Cell Biol* 140: 873–83.

Lindsley, D. L. and Grell, E. H. (1968). *Genetic Variations of Drosophila melanogaster*. Carnegie Institute of Washington Publication 624.

Lindsley, D. L. and Sandler, L. (1963). Construction of the compound-*X* chromosomes in *Drosophila melanogaster* by means of the *Bar Stone* duplication. In: W. J. Burdette (ed.), *Methodology in Basic Genetics*. Holden-Day, San Francisco, pp. 390–403.

Lindsley, D. L. and Sandler, L. (1977). The genetic analysis of meiosis in female *Drosophila melanogaster*. *Philos Trans R Soc Lond, Ser B* 277: 295–312.

Lindsley, D. L. and Zimm, G. C. (1992). *The Genome of Drosophila melanogaster*. Academic Press, San Diego.

Lindsley, D. L., Sandler, L., Baker, B. S., Carpenter, A. T., Denell, R. E., Hall, J. C., Jacobs, P. A., Miklos, G. L., Davis, B. K., Gethmann, R. C., Hardy, R. W., Steven, A. H., Miller, M., Nozawa, H., Parry, D. M., and Gould-Somero,

M. (1972). Segmental aneuploidy and the genetic gross structure of the *Drosophila* genome. *Genetics* 71: 157–84.

Liu, S., McLeod, E., and Jack, J. (1991). Four distinct regulatory regions of the cut locus and their effect on cell type specification in Drosophila. *Genetics* 127: 151–9.

Loidl, J. (1990). The initiation of meiotic chromosome pairing: the cytological view. *Genome* 33: 759–78.

Loidl, J., Scherthan, H., and Kaback, D. B. (1994). Physical association between nonhomologous chromosomes precedes distributive disjunction in yeast. *Proc Natl Acad Sci USA* 91: 331–4.

Lundgren, K., Walworth, N., Booher, R., Dembski, M., Kirschner, M., and Beach, D. (1991). mik1 and wee1 cooperate in the inhibitory tyrosine phosphorylation of cdc2. *Cell* 64: 1111–22.

Lyon, M. F., Rastan, S., Brown, S. D. M., and the International Committee on Standardized Genetic Nomenclature for Mice (eds.) (1995). *Genetic Variants and Strains of the Laboratory Mouse*. Oxford University Press, New York.

Maggert, K. A. and Karpen, G. H. (2001). The activation of a neocentromere in *Drosophila* requires proximity to an endogenous centromere. *Genetics* 158: 1615–28.

Magliery, T. J., Anderson, J. C., and Schultz, P. G. (2001). Expanding the genetic code: selection of efficient suppressors of four-base codons and identification of "shifty" four-base codons with a library approach in *Escherichia coli*. *J Mol Biol* 307: 755–69.

Maguire, M. P., Paredes, A. M., and Riess, R. W. (1991). The desynaptic mutant of maize as a combined defect of synaptonemal complex and chiasma maintenance. *Genome* 34: 879–87.

Mahadevaiah, S. K., Turner, J. M., Baudat, F., Rogakou, E. P., de Boer, P., Blanco-Rodriguez, J., Jasin, M., Keeney, S., Bonner, W. M., and Burgoyne, P. S. (2001). Recombinational DNA double-strand breaks in mice precede synapsis. *Nat Genet* 27: 271–6.

Maine, E. M. (2000). A conserved mechanism for posttranscriptional gene silencing? *Genome Biol* 1: R1018.

Manson, M. D. (2000). Allele-specific suppression as a tool to study protein–protein interactions in bacteria. *Methods* 20: 18–34.

Matthies, H. J., Messina, L. G., Namba, R., Greer, K. J., Walker, M. Y., and Hawley, R. S. (1999). Mutations in the *alpha-tubulin 67C* gene specifically impair achiasmate segregation in *Drosophila melanogaster*. *J Cell Biol* 147: 1137–44.

Matzke, M., Mette, M. F., Jakowitsch, J., Kanno, T., Moscone, E. A., van der Winden, J., and Matzke, A. J. (2001). A test for transvection in plants: DNA pairing may lead to trans-activation or silencing of complex heteroalleles in tobacco. *Genetics* 158: 451–61.

McKee, B. D. (1998). Pairing sites and the role of chromosome pairing in meiosis and spermatogenesis in male *Drosophila*. *Curr Top Dev Biol* 37: 77–115.

McKee, B. D. and Karpen, G. H. (1990). *Drosophila* ribosomal RNA genes function as an *X–Y* pairing site during male meiosis. *Cell* 61: 61–72.

McKee, B. D., Habera, L., and Vrana, J. A. (1992). Evidence that intergenic spacer repeats of *Drosophila melanogaster* rRNA genes function as *X–Y* pairing sites in male meiosis, and a general model for achiasmatic pairing. *Genetics* 132: 529–44.

McKee, B. D., Lumsden, S. E., and Das, S. (1993). The distribution of male meiotic pairing sites on chromosome 2 of *Drosophila melanogaster*: meiotic pairing and segregation of *2-Y* transpositions. *Chromosoma* 102: 180–94.

McKim, K. S. and Hayashi-Hagihara, A. (1998). *mei-W68* in *Drosophila melanogaster* encodes a Spo11 homolog: evidence that the mechanism for initiating meiotic recombination is conserved. *Genes Dev* 12: 2932–42.

McKim, K. S., Heschl, M. F., Rosenbluth, R. E., and Baillie, D. L. (1988a). Genetic organization of the *unc-60* region in *Caenorhabditis elegans*. *Genetics* 118: 49–59.

McKim, K. S., Howell, A. M., and Rose, A. M. (1988b). The effects of translocations on recombination frequency in *Caenorhabditis elegans*. *Genetics* 120: 987–1001.

McKim, K. S., Starr, T., and Rose, A. M. (1992). Genetic and molecular analysis of the dpy-14 region in *Caenorhabditis elegans*. *Mol Gen Genet* 233: 241–51.

McKim, K. S., Peters, K., and Rose, A. M. (1993). Two types of sites required for meiotic chromosome pairing in *Caenorhabditis elegans*. *Genetics* 134: 749–68.

McKim, K. S., Dahmus, J. B., and Hawley, R. S. (1996). Cloning of the Drosophila melanogaster meiotic recombination gene *mei-218*: a genetic and molecular analysis of interval 15E. *Genetics* 144: 215–28.

McKim, K. S., Green-Marroquin, B. L., Sekelsky, J. J., Chin, G., Steinberg, C., Khodosh, R., and Hawley, R. S. (1998). Meiotic synapsis in the absence of recombination. *Science* 279: 876–8.

McLaren, A. (1999). Too late for the midwife toad: stress, variability and Hsp90. *Trends Genet* 15: 169–71.

Mendenhall, M. D., Leeds, P., Fen, H., Mathison, L., Zwick, M., Sleiziz, C., and Culbertson, M. R. (1987). Frameshift suppressor mutations affecting the major glycine transfer RNAs of *Saccharomyces cerevisiae*. *J Mol Biol* 194: 41–58.

Merrill, C., Bayraktaroglu, L., Kusano, A., and Ganetzky, B. (1999). Truncated RanGAP encoded by the *Segregation Distorter* locus of *Drosophila*. *Science* 283: 1742–5.

Meselson, M. S. and Radding, C. M. (1975). A general model for genetic recombination. *Proc Natl Acad Sci USA* 72: 358–61.

Meyerowitz, E. M. and Somerville, C. R. (eds.) (1994). *Arabidopsis*. Cold Spring Harbor Monograph 27, 1300 pp.

Miller, L. M., Waring, D. A., and Kim, S. K. (1996). Mosaic analysis using a ncl-1 (+) extrachromosomal array reveals that lin-31 acts in the Pnp cells during *Caenorhabditis elegans* vulval development. *Genetics* 143: 1181–91.

Moir, D. and Botstein, D. (1982). Determination of the order of gene function in the yeast nuclear division pathway using *cs* and *ts* mutants. *Genetics* 100: 565–77.

Molnar, M., Bahler, J., Kohli, J., and Hiraoka, Y. (2001). Live observation of fission yeast meiosis in recombination-deficient mutants: a study on achiasmate chromosome segregation. *J Cell Sci* 114: 2843–53.

Morris, J. R., Chen, J. L., Geyer, P. K., and Wu, C. T. (1998). Two modes of transvection: enhancer action in trans and bypass of a chromatin insulator in cis. *Proc Natl Acad Sci USA* 95: 10740–5.

Morris, J. R., Chen, J., Filandrinos, S. T., Dunn, R. C., Fisk, R., Geyer, P. K., and Wu, C. (1999a). An analysis of transvection at the *yellow* locus of *Drosophila melanogaster*. *Genetics* 151: 633–51.

Morris, J. R., Geyer, P. K., and Wu, C. T. (1999b). Core promoter elements can regulate transcription on a separate chromosome in *trans*. *Genes Dev* 13: 253–8.

Mounkes, L. C. and Fuller, M. T. (1999). Molecular characterization of mutant alleles of the DNA repair/basal transcription factor haywire/ERCC3 in *Drosophila*. *Genetics* 152: 291–7.

Mounkes, L. C., Jones, R. S., Liang, B. C., Gelbart, W., and Fuller, M. T. (1992). A *Drosophila* model for xeroderma pigmentosum and Cockayne's syndrome: *haywire* encodes the fly homolog of ERCC3, a human excision repair gene. *Cell* 71: 925–37.

Muller, H. J. (1932). Further studies on the nature and causes of gene mutations. *Proc 6th Intl Congr Genet (Ithaca)* I: 213–55.

Murgola, E. J. (1985). tRNA, suppression, and the code. *Ann Rev Genet* 19: 57–80.

Murphy, T. D. and Karpen, G. H. (1995). Localization of centromere function in a *Drosophila* minichromosome. *Cell* 82: 599–609.

Nagel, A. C., Yu, Y., and Preiss, A. (1999). *Enhancer of split [E(spl)(D)]* is a gro-independent, hypermorphic mutation in *Drosophila*. *Dev Genet* 25: 168–79.

Nassif, N., Penney, J., Pal, S., Engels, W. R., and Gloor, G. B. (1994). Efficient copying of nonhomologous sequences from ectopic sites via P-element-induced gap repair. *Mol Cell Biol* 14: 1613–25.

Nelson, C. R. and Szauter, P. (1992). Timing of mitotic chromosome loss caused by the ncd mutation of *Drosophila melanogaster. Cell Motil Cytoskeleton* 23: 34–44.

Nesbitt, M. N. and Gartler, S. M. (1971). The application of genetic mosaicism to developmental problems. *Ann Rev Genet* 5: 143–62.

Nicklas, R. B. (1967). Chromosome micromanipulation. II. Induced reorientation and the experimental control of segregation in meiosis. *Chromosoma* 21: 17–50.

Nicklas, R. B. (1974). Chromosome segregation mechanisms. *Genetics* 78: 205–13.

Nicklas, R. B. (1977). Chromosome distribution: experiments on cell hybrids and in vitro. *Philos Trans R Soc Lond, Ser B* 277: 267–76.

Nicklas, R. B. and Koch, C. A. (1969). Chromosome micromanipulation. 3. Spindle fiber tension and the reorientation of mal-oriented chromosomes. *J Cell Biol* 43: 40–50.

Nicolas, A. and Petes, T. D. (1994). Polarity of meiotic gene conversion in fungi: contrasting views. *Experientia* 50: 242–52.

Nicklas, R. B. and Staehly, C. A. (1967). Chromosome micromanipulation. I. The mechanics of chromosome attachment to the spindle. *Chromosoma* 21: 1–16.

Nicolas, A., Treco, D., Schultes, N. P., and Szostak, J. W. (1989). An initiation site for meiotic gene conversion in the yeast *Saccharomyces cerevisiae. Nature* 338: 35–9.

Nimmo, E. R., Pidoux, A. L., Perry, P. E., and Allshire, R. C. (1998). Defective meiosis in telomere-silencing mutants of *Schizosaccharomyces pombe. Nature* 392: 825–8.

Niwa, O., Matsumoto, T., Chikashige, Y., and Yanagida, M. (1989). Characterization of *Schizosaccharomyces pombe* minichromosome deletion derivatives and a functional allocation of their centromere. *EMBO J* 8: 3045–52.

Nonet, M. L. and Young, R. A. (1989). Intragenic and extragenic suppressors of mutations in the heptapeptide repeat domain of *Saccharomyces cerevisiae* RNA polymerase II. *Genetics* 123: 715–24.

Novitski, E. and Braver, G. (1954). An analysis of crossing-over within a heterozygous inversion *Drosophila melanogaster. Genetics* 39: 197–209.

Nusslein-Volhard, C., Wieschaus, E., and Kluding, H. (1984). Mutations affecting the pattern of the larval cuticle in *Drosophila melanogaster. Roux's Arch Dev Biol* 193: 267–82.

Okagaki, R. J. and Weil, C. F. (1997). Analysis of recombination sites within the maize *waxy* locus. *Genetics* 147: 815–21.

Oliver, P. (1940). A reversion to wildtype associated with crossing over in *Drosophila melanogaster. Proc Natl Acad Sci USA* 26: 452–4.

Oliver, C. P. and Green, M. M. (1944). Heterosis in compounds of lozenge-alleles of *Drosophila melanogaster. Genetics* 29: 331–47.

Orr-Weaver, T. (1996). Meiotic nondisjunction does the two-step. *Nat Genet* 14: 374–6.

Orr-Weaver, T. L., Szostak, J. W., and Rothstein, R. J. (1981). Yeast transformation: a model system for the study of recombination. *Proc Natl Acad Sci USA* 78: 6354–8.

Othman, M. I., Sullivan, S. A., Skuta, G. L., et al. (1998). Autosomal dominant nanophthalmos (NNO1) with high hyperopia and angle-closure glaucoma maps to chromosome 11. *Am J Hum Genet* 63(5): 1411–18.

Ott, J. (2001). Major strengths and weaknesses of the lod score method. *Adv Genet* 42: 125–32.

Page, D. R. and Grossniklaus, U. (2002). The art and design of genetic screens: *Arabidopsis thaliana. Nat Rev Genet* 3: 124–36.

Page, M. F., Carr, B., Anders, K. R., Grimson, A., and Anderson, P. (1999). SMG-2 is a phosphorylated protein required for mRNA surveillance in *Caenorhabditis elegans* and related to Upf1p of yeast. *Mol Cell Biol* 19: 5943–51.

Panchal, R. G., Wang, S., McDermott, J., and Link Jr., C. J. (1999). Partial functional correction of xeroderma pigmentosum group A cells by suppressor tRNA. *Hum Gene Ther* 10: 2209–19.

Paques, F. and Haber, J. E. (1999). Multiple pathways of recombination induced by double-strand breaks in *Saccharomyces cerevisiae. Microbiol Mol Biol Rev* 63: 349–404.

Park, H. S. and Yamamoto, M. T. (1995). The centric region of the *X* chromosome rDNA functions in male meiotic pairing in *Drosophila melanogaster. Chromosoma* 103: 700–7.

Parker, D. R. and Williamson, J. H. (1976). Aberration induction and segregation in oocytes. In: M. Ashburner and E. Novitski (eds.), *The Genetics and Biology of Drosophila.* Vol. 1c. Academic Press, London, pp. 1252–68.

Parrish, S. and Fire, A. (2001). Distinct roles for RDE-1 and RDE-4 during RNA interference in *Caenorhabditis elegans. Rna* 7: 1397–402.

Perkins, D. D. (1955). Tetrads and crossing over. *J Cell Comp Physiol* 45 (Suppl. 2): 119–47.

Perkins, D. D. (1962). Crossing-over and interference in a multiply marked chromosome arm of *Neurospora. Genetics* 47: 1253–74.

Perkins, D. D. (1997). Chromosome rearrangements in *Neurospora* and other filamentous fungi. *Adv Genet* 36: 239–398.

Phillips, J. P. and Forrest, H. S. (1980). Omnochromes and pteridines. In: M. Ashburner and T. R. F. Wright (eds.), *The Genetics and Biology of Drosophila*. Vol. 2d. Academic Press, London, pp. 541–623.

Phillips-Jones, M. K., Hill, L. S., Atkinson, J., and Martin, R. (1995). Context effects on misreading and suppression at UAG codons in human cells. *Mol Cell Biol* 15: 6593–600.

Ponomareff, G., Giordano, H., DeLotto, Y., and DeLotto, R. (2001). Interallelic complementation at the *Drosophila melanogaster gastrulation-defective* locus defines discrete functional domains of the protein. *Genetics* 159: 635–45.

Pontecorvo, G. (1958). *Trends in Genetic Analysis*. Columbia University Press, New York.

Powers, P. A. and Ganetzky, B. (1991). On the components of segregation distortion in *Drosophila melanogaster*. V. Molecular analysis of the *Sd* locus. *Genetics* 129: 133–44.

Prelich, G. (1999). Suppression mechanisms: themes from variations. *Trends Genet* 15: 261–6.

Ptashne, M. (1987). *A Genetic Switch: Gene Control and Phage Lambda*. Cell Press/Blackwell Scientific Publications, Cambridge, MA/Palo Alto, CA.

Ramer, S. W., Elledge, S. J., and Davis, R. W. (1992). Dominant genetics using a yeast genomic library under the control of a strong inducible promoter. *Proc Natl Acad Sci USA* 89: 11589–93.

Rancourt, D. E., Tsuzuki, T., and Capecchi, M. R. (1995). Genetic interaction between hoxb-5 and hoxb-6 is revealed by nonallelic noncomplementation. *Genes Dev* 9: 108–22.

Rand, J. B. (1989). Genetic analysis of the cha-1-unc-17 gene complex in *Caenorhabditis*. *Genetics* 122: 73–80.

Rasooly, R. S., New, C. M., Zhang, P., Hawley, R. S., and Baker, B. S. (1991). The *lethal(1)TW-6cs* mutation of *Drosophila melanogaster* is a dominant antimorphic allele of *nod* and is associated with a single base change in the putative ATP-binding domain. *Genetics* 129: 409–22.

Rawls, J. M. and Porter, L. A. (1979). Organization of the rudimentary wing locus in *Drosophila melanogaster*. *Genetics* 93: 143–61.

Raz, E., Schejter, E. D., and Shilo, B. Z. (1991). Interallelic complementation among DER/flb alleles: implications for the mechanism of signal transduction by receptor-tyrosine kinases. *Genetics* 129: 191–201.

Rebay, I., Chen, F., Hsiao, F., Kolodziej, P. A., Kuang, B. H., Laverty, T., Suh, C., Voas, M., Williams, A., and Rubin, G. M. (2000). A genetic screen for novel components of the Ras/Mitogen-activated protein kinase signaling pathway that interact with the *yan* gene of *Drosophila* identifies split ends, a new RNA recognition motif-containing protein. *Genetics* 154: 695–712.

Regan, C. L. and Fuller, M. T. (1988). Interacting genes that affect microtubule function: the nc2 allele of the *haywire* locus fails to complement mutations in the testis-specific beta-tubulin gene of *Drosophila*. *Genes Dev* 2: 82–92.

Reich, D. E., Cargill, M., Bolk, S., Ireland, J., Sabeti, P. C., Richter, D. J., Lavery, T., Kouyoumjian, R., Farhadian, S. F., Ward, R., and Lander, E. S. (2001). Linkage disequilibrium in the human genome. *Nature* 411: 199–204.

Reinke, R. and Zipursky, S. L. (1988). Cell–cell interaction in the *Drosophila* retina: the bride of sevenless gene is required in photoreceptor cell R8 for R7 cell development. *Cell* 55: 321–30.

Ren, X., Eisenhour, L., Hong, C., Lee, Y., and McKee, B. D. (1997). Roles of rDNA spacer and transcription unit-sequences in X–Y meiotic chromosome pairing in *Drosophila melanogaster* males. *Chromosoma* 106: 29–36.

Rice, J. P., Saccone, N. L., and Corbett, J. (2001). The lod score method. *Adv Genet* 42: 99–113.

Riddle, D. L., Blumenthal, T., Meyer, B. J., and Priess, J. R. (1997). *C. Elegans II*. Cold Spring Harbor Press, New York.

Robinson, D. N. and Cooley, L. (1997). Examination of the function of two kelch proteins generated by stop codon suppression. *Development* 124: 1405–17.

Rockmill, B. and Roeder, G. S. (1990). Meiosis in asynaptic yeast. *Genetics* 126: 563–74.

Rockmill, B. and Roeder, G. S. (1998). Telomere-mediated chromosome pairing during meiosis in budding yeast. *Genes Dev* 12: 2574–86.

Roeder, G. S. (1997). Meiotic chromosomes: it takes two to tango. *Genes Dev* 11: 2600–21.

Romanienko, P. J. and Camerini-Otero, R. D. (2000). The mouse *Spo11* gene is required for meiotic chromosome synapsis. *Mol Cell* 6: 975–87.

Rorth, P., Szabo, K., Bailey, A., Laverty, T., Rehm, J., Rubin, G. M., Weigmann, K., Milan, M., Benes, V., Ansorge, W., and Cohen, S. M. (1998). Systematic gain-of-function genetics in *Drosophila*. *Development* 125: 1049–57.

Rosen, C., Dorsett, D., and Jack, J. (1998). A proline-rich region in the Zeste protein essential for transvection and white repression by Zeste. *Genetics* 148: 1865–74.

Rosenbluth, R. E. and Baillie, D. L. (1981). The genetic analysis of a reciprocal translocation, *eT1(III; V)*, in *Caenorhabditis elegans*. *Genetics* 99: 415–28.

Ross, L. O., Maxfield, R., and Dawson, D. (1996a). Exchanges are not equally able to enhance meiotic chromosome segregation in yeast. *Proc Natl Acad Sci USA* 93: 4979–83.

Ross, L. O., Rankin, S., Shuster, M. F., and Dawson, D. S. (1996b). Effects of homology, size and exchange of the meiotic segregation of model chromosomes in *Saccharomyces cerevisiae*. *Genetics* 142: 79–89.

Rossignol, J. L., Paquette, N., and Nicolas, A. (1978). Aberrant 4+:4− asci, disparity in the direction of gene conversion, and frequencies of conversion in *Ascobolous immersus*. *Cold Spring Harbor Symp Quant Biol* 43: 1343–56.

Rossignol, J. L., Nicolas, A., Hamza, H., and Langin, T. (1984). Origins of gene conversion and reciprocal exchange in *Ascobolus*. *Cold Spring Harbor Symp Quant Biol* 49: 13–21.

Russell, S. (2000). The Drosophila dominant wing mutation Dichaete results from ectopic expression of a Sox-domain gene. *Mol Gen Genet* 263: 690–701.

Rutherford, S. L. and Lindquist, S. (1998). Hsp90 as a capacitor for morphological evolution. *Nature* 396: 336–42.

Ruvkun, G., Ambros, V., Coulson, A., Waterston, R., Sulston, J., and Horvitz, H. R. (1989). Molecular genetics of the *Caenorhabditis elegans* heterochronic gene *lin-14*. *Genetics* 121: 501–16.

Samson, M. L., Lisbin, M. J., and White, K. (1995). Two distinct temperature-sensitive alleles at the elav locus of *Drosophila* are suppressed nonsense mutations of the same tryptophan codon. *Genetics* 141: 1101–11.

Sanchez Moran, E., Armstrong, S. J., Santos, J. L., Franklin, F. C. and Jones, G. H. (2001). Chiasma formation in *Arabidopsis thaliana* accession Wassileskija and in two meiotic mutants. *Chromosome Res* 9: 121–8.

Sandler, L., Lindsley, D. L., Nicoletti, B., and Trippa, G. (1968). Mutants affecting meiosis in natural populations of *Drosophila melanogaster*. *Genetics* 60: 525–58.

Sandrock, T. M., O'Dell, J. L., and Adams, A. E. (1997). Allele-specific suppression by formation of new protein–protein interactions in yeast. *Genetics* 147: 1635–42.

Sandrock, T. M., Brower, S. M., Toenjes, K. A., and Adams, A. E. (1999). Suppressor analysis of fimbrin (Sac6p) overexpression in yeast. *Genetics* 151: 1287–97

Saunders, W., Lengyel, V., and Hoyt, M. A. (1997). Mitotic spindle function in *Saccharomyces cerevisiae* requires a balance between different types of kinesin-related motors. *Mol Biol Cell* 8: 1025–33.

Scherthan, H. (1997). Chromosome behavior in earliest meiotic prophase. *Chromosomes Today* 12: 217–48.

Schmidt, A. and Hall, M. N. (1998). Signaling to the actin cytoskeleton. *Ann Rev Cell Dev Biol* 14: 305–38.

Schmuckli-Maurer, J. and Heyer, W. D. (2000). Meiotic recombination in RAD54 mutants of *Saccharomyces cerevisiae*. *Chromosoma* 109: 86–93.

Schwacha, A. and Kleckner, N. (1994). Identification of joint molecules that form frequently between homologs but rarely between sister chromatids during yeast meiosis. *Cell* 76: 51–63.

Schwacha, A. and Kleckner, N. (1995). Identification of double Holliday junctions as intermediates in meiotic recombination. *Cell* 83: 783–91.

Schwacha, A. and Kleckner, N. (1997). Interhomolog bias during meiotic recombination: meiotic functions promote a highly differentiated interhomolog-only pathway. *Cell* 90: 1123–35.

Scott, K. C., Taubman, A. D., and Geyer, P. K. (1999). Enhancer blocking by the *Drosophila gypsy* insulator depends upon insulator anatomy and enhancer strength. *Genetics* 153: 787–98.

Searles, L. L. and Voelker, R. A. (1986). Molecular characterization of the Drosophila *vermilion* locus and its suppressible alleles. *Proc Natl Acad Sci USA* 83: 404–8.

Segraves, W. A., Louis, C., Tsubota, S., Schedl, P., Rawls, J. M., and Jarry, B. P. (1984). The *rudimentary* locus of *Drosophila melanogaster*. *J Mol Biol* 175: 1–17.

Sekelsky, J. J., McKim, K. S., Chin, G. M., and Hawley, R. S. (1995). The *Drosophila* meiotic recombination gene *mei-9* encodes a homologue of the yeast excision repair protein Rad1. *Genetics* 141: 619–27.

Seydoux, G. and Greenwald, I. (1989). Cell autonomy of lin-12 function in a cell fate decision in *C. elegans*. *Cell* 57: 1237–45.

Sharp, D. J., Rogers, G. C., and Scholey, J. M. (2000). Microtubule motors in mitosis. *Nature* 407: 41–7.

Sherman, F. and Helms, C. (1978). A chromosomal translocation causing overproduction of iso-2-cytochrome c in yeast. *Genetics* 88: 689–707.

Simon, M. A., Bowtell, D. D., Dodson, G. S., Laverty, T. R., and Rubin, G. M. (1991). Ras1 and a putative guanine nucleotide exchange factor perform crucial steps in signaling by the sevenless protein tyrosine kinase. *Cell* 67: 701–16.

Simpson, L. and Wieschaus, E. (1990). Zygotic activity of the *nullo* locus is required to stabilize the actin-myosin network during cellularization in *Drosophila*. *Development* 110: 851–63.

Smolik-Utlaut, S. M. and Gelbart, W. M. (1987). The effects of chromosomal rearrangements on the *zeste–white* interaction in Drosophila melanogaster. *Genetics* 116: 285–98.

Stadler, D. R. and Towe, A. M. (1963). Recombination of allelic cysteine mutants in *Neurospora*. *Genetics* 48: 1323–44.

Stahl, F. W. (1969). *The Mechanics of Inheritance*. Prentice-Hall, Englewood Cliffs, NJ.

Stahl, F. W. (1994). The Holliday junction on its thirtieth anniversary. *Genetics* 138: 241–6.

Stahl, F. (1996). Meiotic recombination in yeast: coronation of the double-strand-break repair model. *Cell* 87: 965–68.

Stanford, W. L., Cohn, J. B., and Cordes, S. P. (2001). Gene-trap mutagenesis: past, present and beyond. *Nat Rev Genet* 2: 756–68.

Stearns, T. and Botstein, D. (1988). Unlinked noncomplementation: isolation of new conditional-lethal mutations in each of the tubulin genes of *Saccharomyces cerevisiae*. *Genetics* 119: 249–60.

Steiner, N. C. and Clarke, L. (1994). A novel epigenetic effect can alter centromere function in fission yeast. *Cell* 79: 865–74.

Stern, C. (1936). Somatic crossing over and segregation in Drosophila melanogaster. *Genetics* 21: 625–730.

Stern, C. (1968). Genetic mosaics in animals and man. In: *Genetic Mosaics and other Essays*. Harvard University Press, Cambridge, MA, pp. 130–73.

Strathern, J. N., Klar, A. J., Hicks, J. B., Abraham, J. A., Ivy, J. M., Nasmyth, K. A., and McGill, C. (1982). Homothallic switching of yeast mating type cassettes is initiated by a double-stranded cut in the MAT locus. *Cell* 31: 183–92.

Streisinger, G., Singer, F., Walker, C., Knauber, D., and Dower, N. (1986). Segregation analyses and gene–centromere distances in zebrafish. *Genetics* 112: 311–19.

Sturtevant, A. H. and Beadle, G. (1936). The relationships of inversions in the X chromosome of *Drosophila melanogaster*. *Genetics* 21: 554–604.

Sullivan, W., Ashburner, M., and Hawley, R. S. (2000). *Drosophila Protocols*. Cold Spring Harbor Press, New York.

Sun, H., Treco, D., Schultes, N. P., and Szostak, J. W. (1989). Double-strand breaks at an initiation site for meiotic gene conversion. *Nature* 338: 87–90.

Sun, X., Wahlstrom, J., and Karpen, G. (1997). Molecular structure of a functional *Drosophila* centromere. *Cell* 91: 1007–19.

Svoboda, P., Stein, P., and Schultz, R. M. (2001). RNAi in mouse oocytes and preimplantation embryos: effectiveness of hairpin dsRNA. *Biochem Biophys Res Commun* 287: 1099–104.

Szauter, P. (1984). An analysis of regional constraints on exchange in *Drosophila melanogaster* using recombination-defective meiotic mutants. *Genetics* 106: 45–71.

Szostak, J. W., Orr-Weaver, T. L., Rothstein, R. J., and Stahl, F. W. (1983). The double-strand-break repair model for recombination. *Cell* 33: 25–35.

Tang, T. T., Bickel, S. E., Young, L. M., and Orr-Weaver, T. L. (1998). Maintenance of sister-chromatid cohesion at the centromere by the *Drosophila* MEI-S332 protein. *Genes Dev* 12: 3843–56.

Tartof, K. D. and Henikoff, S. (1991). Trans-sensing effects from Drosophila to humans. *Cell* 65: 201–3.

Taylor, E. W., Xu, J., Jabs, E. W., and Meyers, D. A. (1997). Linkage analysis of genetic disorders. *Methods Mol Biol* 68: 11–25.

Theurkauf, W. E. and Hawley, R. S. (1992). Meiotic spindle assembly in Drosophila females: behavior of nonexchange chromosomes and the effects of mutations in the nod kinesin-like protein. *J Cell Biol* 116: 1167–80.

Thompson, C. M., Koleske, A. J., Chao, D. M., and Young, R. A. (1993). A multisubunit complex associated with the RNA polymerase II CTD and TATA-binding protein in yeast. *Cell* 73: 1361–75.

Timmons, L., Court, D. L., and Fire, A. (2001). Ingestion of bacterially expressed dsRNAs can produce specific and potent genetic interference in *Caenorhabditis elegans*. *Gene* 263: 103–12.

Tomkiel, J. E., Wakimoto, B. T., and Briscoe Jr., A. (2001). The *teflon* gene is required for maintenance of autosomal homolog pairing at meiosis I in male *Drosophila melanogaster*. *Genetics* 157: 273–81.

Trelles-Sticken, E., Loidl, J., and Scherthan, H. (1999). Bouquet formation in budding yeast: initiation of recombination is not required for meiotic telomere clustering. *J Cell Sci* 112 (Pt 5): 651–8.

Tsai, S. F., Jang, C. C., Prikhod'ko, G. G., Bessarab, D. A., Tang, C. Y., Pflugfelder, G. O., and Sun, Y. H. (1997). Gypsy retrotransposon as a tool for the in vivo analysis of the regulatory region of the optomotor-blind gene in Drosophila. *Proc Natl Acad Sci USA* 94: 3837–41.

Tsubota, S. I. and Fristrom, J. W. (1981). Genetic and biochemical properties of revertants at the rudimentary locus in *Drosophila melanogaster*. *Mol Gen Genet* 183: 270–6.

van Heemst, D. and Heyting, C. (2000). Sister chromatid cohesion and recombination in meiosis. *Chromosoma* 109: 10–26.

Vazquez, J., Belmont, A., and Sedat, J. (2002). The dynamics of homologous pairing during male meiosis in Drosophila. *Curr Biol* (in press).

Villeneuve, A. M. (1994). A cis-acting locus that promotes crossing over between X chromosomes in *Caenorhabditis elegans*. *Genetics* 136: 887–902.

Villeneuve, A. M. and Meyer, B. J. (1990a). The role of sdc-1 in the sex determination and dosage compensation decisions in *Caenorhabditis elegans*. *Genetics* 124: 91–114.

Villeneuve, A. M. and Meyer, B. J. (1990b). The regulatory hierarchy controlling sex determination and dosage compensation in Caenorhabditis elegans. *Adv Genet* 27: 117–88.

Vinh, D. B., Welch, M. D., Corsi, A. K., Wertman, K. F., and Drubin, D. G. (1993). Genetic evidence for functional interactions between actin noncomplementing (Anc) gene products and actin cytoskeletal proteins in *Saccharomyces cerevisiae*. *Genetics* 135: 275–86.

Wagner Jr., R. and Meselson, M. (1976). Repair tracts in mismatched DNA heteroduplexes. *Proc Natl Acad Sci USA* 73: 4135–9.

Walker, M. Y. and Hawley, R. S. (2000). Hanging on to your homolog: the roles of pairing, synapsis and recombination in the maintenance of homolog adhesion. *Chromosoma* 109: 3–9.

Washburn, T. and O'Tousa, J. E. (1992). Nonsense suppression of the major rhodopsin gene of *Drosophila*. *Genetics* 130: 585–95.

Waterhouse, P. M., Graham, M. W., and Wang, M. B. (1998). Virus resistance and gene silencing in plants can be induced by simultaneous expression of sense and antisense RNA. *Proc Natl Acad Sci USA* 95: 13959–64.

Watson, J. D. and Crick, F. H. (1953). Molecular structure of nucleic acids: a structure for deoxyribose nucleic acid. *Nature* 171: 737–8.

Weigel, D. and Glazebrook, J. (2002). *Arabidopsis: A Laboratory Manual*. Cold Spring Harbor Laboratory Press, New York.

Weinstein, A. (1932). A theoretical and experimental analysis of crossing over. *Proc Sixth Intl Congr Genet* 2: 206–8.

Weiss, K. M. and Clark, A. G. (2002). Linkage disequilibrium and the mapping of complex human traits. *Trends Genet* 18: 19–24.

Welch, M. D. and Drubin, D. G. (1994). A nuclear protein with sequence similarity to proteins implicated in human acute leukemias is important for cellular morphogenesis and actin cytoskeletal function in *Saccharomyces cerevisiae*. *Mol Biol Cell* 5: 617–32.

Welch, M. D., Vinh, D. B., Okamura, H. H., and Drubin, D. G. (1993). Screens for extragenic mutations that fail to complement act1 alleles identify genes that are important for actin function in *Saccharomyces cerevisiae*. *Genetics* 135: 265–74.

Westerfield, M. (2000). *The Zebrafish Book*. University of Oregon Press, Eugene, OR.

Wettstein, D. von (1984). The synaptonemal complex and genetic segregation. In: C. W. Evans and H. G. Dickinson (eds.), *Controlling Events in Meiosis*. The Company of Biologists, Cambridge, pp. 195–232.

White, C. I. and Haber, J. E. (1990). Intermediates of recombination during mating type switching in *Saccharomyces cerevisiae*. *EMBO J* 9: 663–73.

Whitehouse, H. K. (1982). *Genetic Recombination: Understanding the Mechanism*. Wiley & Sons, New York.

Whyte, W. L., Irick, H., Arbel, T., Yasuda, G., French, R. L., Falk, D. R., and Hawley, R. S. (1993). The genetic analysis

of achiasmate segregation in *Drosophila melanogaster*. III. The wild-type product of the Axs gene is required for the meiotic segregation of achiasmate homologs. *Genetics* 134: 825–35.

Wieschaus, E. and Gehring, W. (1976). Clonal analysis of primordial disc cells in the early embryo of *Drosophila melanogaster*. *Dev Biol* 50: 249–63.

Wildenberg, J. and Meselson, M. (1975). Mismatch repair in heteroduplex DNA. *Proc Natl Acad Sci USA* 72: 2202–6.

Williams, B. C., Murphy, T. D., Goldberg, M. L., and Karpen, G. H. (1998). Neocentromere activity of structurally acentric mini-chromosomes in *Drosophila*. *Nat Genet* 18: 30–37.

Willins, D. A., Xiang, X., and Morris, N. R. (1995). An alpha-tubulin mutation suppresses nuclear migration mutations in *Aspergillus nidulans*. *Genetics* 141: 1287–98.

Wills, N., Gesteland, R. F., Karn, J., Barnett, L., Bolten, S., and Waterston, R. H. (1983). The genes *sup-7 X* and *sup-5 III* of *C. elegans* suppress amber nonsense mutations via altered transfer RNA. *Cell* 33: 575–83.

Winnier, G. E., Kume, T., Deng, K., Rogers, R., Bundy, J., Raines, C., Walter, M. A., Hogan, B. L., and Conway, S. J. (1999). Roles for the winged helix transcription factors MF1 and MFH1 in cardiovascular development revealed by nonallelic noncomplementation of null alleles. *Dev Biol* 213: 418–31.

Wolf, K. W. (1994). How meiotic cells deal with non-exchange chromosomes. *Bioessays* 16: 107–14.

Wu, C. I., Lyttle, T. W., Wu, M. L., and Lin, G. F. (1988). Association between a satellite DNA sequence and the *Responder of Segregation Distorter* in *D. melanogaster*. *Cell* 54: 179–89.

Wu, T. C. and Lichten, M. (1994). Meiosis-induced double-strand break sites determined by yeast chromatin structure. *Science* 263: 515–18.

Wu, T. C. and Lichten, M. (1995). Factors that affect the location and frequency of meiosis-induced double-strand breaks in *Saccharomyces cerevisiae*. *Genetics* 140: 55–66.

Wu, C. T. and Morris, J. R. (1999). Transvection and other homology effects. *Curr Opin Genet Dev* 9: 237–46.

Yang, S., Tutton, S., Pierce, E., and Yoon, K. (2001). Specific double-stranded RNA interference in undifferentiated mouse embryonic stem cells. *Mol Cell Biol* 21: 7807–16.

Yook, K. J., Proulx, S. R., and Jorgensen, E. M. (2001). Rules of nonallelic noncomplementation at the synapse in *Caenorhabditis elegans*. *Genetics* 158: 209–20.

Yu, Y. and Bradley, A. (2001). Engineering chromosomal rearrangements in mice. *Nat Rev Genet* 2: 780–90.

Zalokar, M., Erk, I., and Santamaria, P. (1980). Distribution of ring-X chromosomes in the blastoderm of gynandromorphic *D. melanogaster*. *Cell* 19: 133–41.

Zhang, P. and Hawley, R. S. (1990). The genetic analysis of distributive segregation in *Drosophila melanogaster*. II. Further genetic analysis of the *nod* locus. *Genetics* 125: 115–27.

Zhao, H. and Speed, T. P. (1998a). Statistical analysis of ordered tetrads. *Genetics* 150: 459–72.

Zhao, H. and Speed, T. P. (1998b). Statistical analysis of half-tetrads. *Genetics* 150: 473–85.

Zhao, H., McPeek, M. S., and Speed, T. P. (1995). Statistical analysis of chromatid interference. *Genetics* 139: 1057–65.

Zheng, B., Sage, M., Cai, W. W., Thompson, D. M., Tavsanli, B. C., Cheah, Y. C., and Bradley, A. (1999). Engineering a mouse balancer chromosome. *Nat Genet* 22: 375–8.

Zhou, J., Lloyd, S. A., and Blair, D. F. (1998). Electrostatic interactions between rotor and stator in the bacterial flagellar motor. *Proc Natl Acad Sci USA* 95: 6436–41.

Zickler, D. and Kleckner, N. (1998). The leptotene–zygotene transition of meiosis. *Ann Rev Genet* 32: 619–97.

Zickler, D. and Kleckner, N. (1999). Meiotic chromosomes: integrating structure and function. *Ann Rev Genet* 33: 603–754.

Zwick, M. E., Cutler, D. J., and Langley, C. H. (1999). Classic Weinstein: tetrad analysis, genetic variation and achiasmate segregation in *Drosophila* and humans. *Genetics* 152: 1615–29.

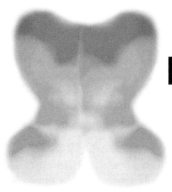

Partial author index

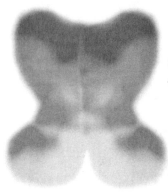

Subject index